Multiple Criteria Decision Support in Engineering Design

Springer
London
Berlin
Heidelberg
New York
Barcelona
Budapest
Hong Kong
Milan
Paris
Santa Clara
Singapore
Tokyo

Pratyush Sen and Jian-Bo Yang

Multiple Criteria Decision Support in Engineering Design

With 83 Figures

Springer

Professor Pratyush Sen
Department of Marine Technology, Armstrong Building, University of Newcastle, Newcastle-upon-Tyne, NE1 7RU, UK

Dr. Jian-Bo Yang
Manchester School of Management, UMIST PO BOX, Manchester, M60 1QD, UK

ISBN-13: 978-1-4471-3022-2 e-ISBN-13: 978-1-4471-3020-8
DOI: 10.1007/978-1-4471-3020-8

British Library Cataloguing in Publication Data
Sen, Pratyush
Multiple criteria decision support in engineering design
1.Engineering design 2.Decision-making
I.Title II.Yang, Jian-Bo
620'.0042

Library of Congress Cataloging-in-Publication Data
Sen, Pratyush, 1948-
Multiple criteria decision support in engineering design /
Pratyush Sen and Jian-Bo Yang.
p. cm.

1. Engineering design. 2. Decision support systems. I. Yang, Jian-Bo, 1961- II. Title.
TA174.S463 1998 98-26520
620'.0042--dc21 CIP

Softcover reprint of the hardcover 1st edition 1995

Typesetting: Camera ready by authors

69/3830-543210 Printed on acid-free paper

Dedicated to Dong-Ling, Indi, Lin and Sree for all their support whilst this book was taking shape.

FOREWORD

This book is an important step forward towards making design an objective process in which decisions can be rationally accounted for. That does not mean that the creativity, the skill, the humanity, and the challenging fun of designing is to be lost. It does mean that design decision makers can be more confident that their design solutions have been well matched to the needs and contraints which represent the ambitions of all their potential customers and, indeed, any who will come into contact with the resulting product.

When designing mighty systems such as large commercial aircraft, for example, it is easy to feel a sense of awe that people can be transported swiftly over large distances in comfort and safety in a massive structure depending on the most demanding technology. It is also easy to forget that it is a flying compromise in many dimensions. It must satisfy (or, preferably, delight) the passengers, the flight crew, the cabin crew, the baggage handlers, the maintainers, the airline which owns it or the leasing agency which owns it. Air trafffic controllers will want it to fit well into their expectations. Those living near the airports it uses will have an interest in its noise and smell. Society at large will have demands and feelings which will extend over fuel consumption, safety, ease of disposal... and so on. All this is too important to be left to the unprovable decisions of specialists working, quite possibly, in isolation from one another. We have moved into an era when all the conflicts and compromises in such an extensive design task, involving huge technical and human systems, need sound, analytical balancing, rendering the decisions effective, logical, and traceable... in an economically short time.

For some ten years I have been fascinated by the way in which Professor Pratyush Sen has taken his constructive and probing approach to Multiple Criteria Decision Making (MCDM) as a means of handling these problems. It is a major component of the work of the highly successful Engineering Design Centre at the University of Newcastle. In my roles as Design Co-ordinator for the Science and Engineering Research Council and, later, as Design Consultant to the Engineering and Physical Sciences Research Council I have been privileged to witness Professor Sen and his group develop, extend, and implement the ideas. He has much to tell us; and there is still more to come as he continues his researches and applies them across the whole industrial spectrum.

Those of us who have practised engineering design know only too well that designing is about trying to obtain the best solution to the problem, *taking everything into consideration.* In other words, to design is to optimise. In this book Professor Sen and his worthy co-author Dr.J.B.Yang have clearly laid out the procedures to do just that. It is an admirable text to help designers. Its approach takes design into the realms of managerial decision making in a way most of us could not have dreamed possible only a few years ago. This, for me, really is Computer Aided Design.

Professor Peter Hills
President, Institution of Engineering Designers
June 1998

Preface

Decision making, in general, and engineering decision making, in particular, often involve the balancing of multiple, potentially conflicting requirements. Classical optimisation deals with these problems by taking the most important requirement as the objective function and the remainder as constraints.

This still leaves the problem of potentially irreconcilable requirements and the usual approach employed to solve this problem is to relax the thresholds of the constraints until feasible solutions emerge. However, there is an alternative body of methods that take a different view. These deal with multiple criteria problems as they appear and employ a range of processes that clarify the consequences of the underlying trade-offs between criteria in configuring alternative solutions. The aim is to use the conflict resolution process as a creative activity. It is in this context that this book has been written. However, two fortunate events have contributed to its actual writing.

Firstly, the setting up of the Engineering Design Centre (EDC) at Newcastle University in 1990 with funding from the Engineering and Physical Sciences Research Council (EPSRC) allowed the first author to assemble a group of researchers under him to continue to more rigorously pursue his personal research interests in the important area of Multiple Criteria Decision Making (MCDM) in engineering design. This was as a part of the overall research portfolio of the EDC. Dr.Yang, the co-author of this book, joined the research group on MCDM in 1991 and worked within this group until 1995 when he left the Newcastle EDC. By the time he left he had made an impressive contribution to the development of the methodological base that is reported in Chapters 3 and 4. Although both authors had a strong research interest in MCDM before the formation of the EDC, the four and a half years together helped to produce a body of work that has proved to be of wide applicability both within the EDC and without.

The second fortunate set of circumstances that has helped this book is the issue of topicality. MCDM as a formal body of methodology has evolved into a discipline in its own right over the last quarter of a century and has now reached a level of maturity that merits its formal adoption in many decision making situations. This provided the technical impetus for sharing some of our experiences with potential users.

Given the personal backgrounds of the authors and the context of the EDC it seemed natural to develop the methodological base as a decision support environment primarily for engineering design. The examples of engineering design decision making used in the text are often drawn from the domain of marine technology which is the base department of the first author. They have been presented in a sufficiently generic manner, however, so as not to pose any difficulties for the average reader, we hope.

The production of the work has also benefited from direct and indirect contributions from several students and research associates from the Decision Support Group at the EDC which the first author leads. In particular thanks are due to David Todd, Tri Achmadi, Raj Subramani, Zhengfu Rao, Peter Meldrum and Jaime Scott. David, in particular, has assisted with the production aspects of all the technical chapters and with technical material for Chapter 5. Both authors also wish to acknowledge the numerous technical discussions held with a large number of interested users of the methodologies. They are too numerous to name but it is a pleasure to acknowledge that their implicit contributions have helped to clarify some of our own thinking in several of the areas. The authors are also grateful for the excellent word processing and editing assistance from Kathleen Heads.

Finally, both authors would like to thank their families for the forbearance, understanding and support during the anti-social hours that had to be kept at times during the writing of this book.

Pratyush Sen
Department of Marine Technology
University of Newcastle

Jian-Bo Yang
School of Management
University of Manchester Institute of
Science & Technology

Table of Contents

1
Introduction

Decision making of all kinds involves the choice of one or more alternatives from a list of options. The list of options would normally all be more or less acceptable solutions for the problem at hand and consequences, both good and bad, flow from the exercise of choice. The aim of rational decision making, therefore, is to maximise the positive consequences and minimise the negative ones. As these consequences are directly related to the decision made or option chosen, it is not unreasonable to treat the consequences as aspects of performance. The decision problem then becomes a matter of considering these aspects of performance of all the options available simultaneously so that the decision maker (DM) can exercise his choice. In other words, rational decision making involves choice within the context of multiple measures of performance or *multiple criteria.*

What then are the basic features of this decision making process?

1.1 What is Multiple Criteria Decision Making (MCDM)?

Most readers of this text will be familiar with multiple criteria problems because such problems arise naturally in many real life situations. However since this text is largely directed at the technical decision-maker it would be rational to begin by answering this simple but specific question. This is because such a decision-maker would need to examine the characteristics of the tools in question before deciding how, or indeed whether, such tools have anything to offer.

In all branches of human endeavour solutions are sought by matching the specific characteristics of the candidate or evolving actions with the specified performance requirements for the situation in hand. Thus a doctor would examine the beneficial and harmful effects of a particular therapy and compare it to others before deciding to put it into effect. Similarly a planning authority would examine all the effects of a particular development before sanctioning a go-ahead or deciding to turn the application down. Even in our personal lives we are continually being confronted with situations that require us to examine alternative actions each of which may be attractive in some respects but with its own drawbacks as well. The common factors running through all of the above situations are as follows:

(i) there is a range of possible actions.
(ii) each action is characterised by a set of consequences some of which are beneficial and others less so.
(iii) the decision maker is required to weigh up the pros and cons before arriving at a preferred action, and to do this he might use a range of decision rules.

The above situations can thus be characterised by the need to make decisions or choices on the basis of a set of actions that have multiple, potentially conflicting performance criteria associated with them. The process by which such decisions are made may be based on subjective reasoning or objective analysis and evaluation but the essential nature of the problem does depend very much on how the inherent conflicts are resolved. It is intuitively clear that different DMs and decision situations would require different conflict resolution strategies so that the action ultimately chosen is determined by a combination of examination of the alternative actions available and the encapsulation of the priority ordering of the DM. This latter factor is particularly important because priority ordering is a natural method of reconciling conflicting requirements. This is because when we are in situations where all that we wish to achieve is not attainable then the rational thing to do would be to revise our expectations or re-order our priorities or do a bit of both. This is the "satisficing" principle of Simon [Simon, 1981] where compromise is necessary if a strict optimisation is not possible.

As priorities are specific to individuals or organisations the "optimal" action therefore becomes personalised. This is quite different from classical optimisation where there is only a single criterion and a set of satisfiable constraints. In classical decision making approaches within the mono-criterion paradigm the formulation of the problem leads naturally to the solution. Once the criterion of interest is agreed upon – cost, for example – the choice of the most attractive action is not a matter of opinion. There can be some argument as to how cost may be computed but in any meaningful problem the method of computation is obviously a part of the definition of the criterion itself. In other words the choice of criterion leads directly to the solution in the mono-criterion paradigm and it is a solution that all parties can agree with. However, as already seen, this is not the case for the multiple criteria formulation as alternative decision makers can and will, in general, have different value systems leading to different priority orderings of the multiple, potentially conflicting performance criteria.

1.2 Relevance of MCDM to Engineering Design

1.2.1 The Structure of a Design Problem.

Engineering design is the process of configuring an artefact so that the performance attributes of the chosen solution meet some functional requirements.

These requirements are usually stated in a specification document of some kind. This often leads to the linear model of design [Pahl 1977] where the solution process is modelled as one that passes through several stages to produce an acceptable design as Figure 1.1 shows.

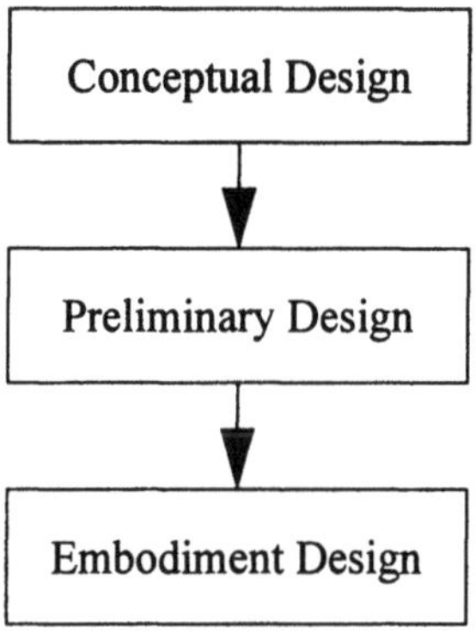

Figure 1.1 *Linear model of design process*

The above structure of the engineering design process is a useful framework for discussion. However, most designers recognise that very often in real design situations the process is not linear but iterative in that developments in one area or level of design alters the freedom of action in another. Moreover, design actions can lead to modifications in the specification. In other words, specification of performance is often affected by an examination of what is technically feasible. This phenomenon can sometimes be the cause of a lot of misunderstanding and argument between a client and contractor but within any given situation such creative rewriting of the specification is one means by which companies remain competitive as it allows them to take advantage of appropriate emerging developments. Since a specification document is, in essence, an elaborate list of requirements, the changing of the specification is tantamount to a virtual re-ordering of technical priorities. This is implicitly a recognition of the multiple criteria nature of design development.

An alternative construct often used in the domain of marine technology, but is of general applicability, is the concept of the design spiral as shown in Figure1.2 [Andrews, 1981].

In this construct the design is seen to progress over time by a series of adjustments until some agreed performance criteria have been met. In other words such a construct usually assumes that all of the design requirements can be met.

However, this may not always be possible unless at least some of the requirements are stated in terms of open-ended statements like "design the lightest structure possible". It is thus implicit that adjustments in the stated requirements have to be made if the currently stipulated thresholds are not all achievable. As already observed, any adjustment of requirements is implicitly a change of priorities. The multiple criteria paradigm makes all such adjustments transparent and hence more capable of adjustment.

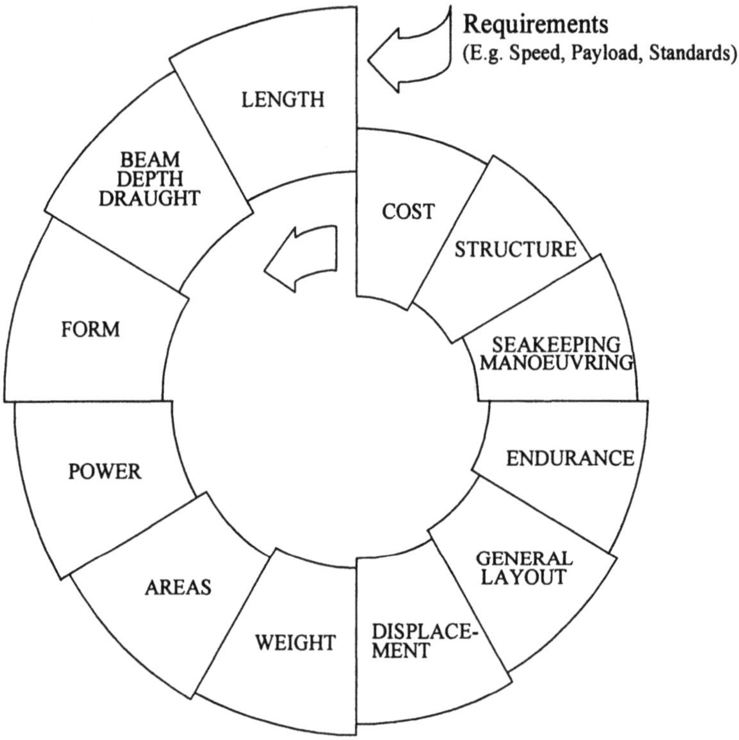

Figure 1.2 *The Design Spiral*

In general, therefore, both the mono-criterion and the multiple criteria optimisation approaches recognise the need for compromise but pursue this in rather different ways. In the latter the compromise is essentially part of the process of decision making and is therefore explicitly catered for in the range of mathematical tools available to deal with such problems. It also implies the presence of mechanisms for capturing the DM's priorities.

However, whether a mono-criterion or a multiple criterion approach is taken to design, both optimisation approaches to design have one common feature. This is that the different design requirements are all simultaneously taken into account. In that sense both approaches are in the spirit of concurrency. This has a direct bearing on issues of technical trade-off, as will be seen later in Chapters 3, 4 and 5.

1.2.2 The Principal Issues in Multiple Criteria Decision Making

In the discussion above it has been implied that multiple criteria decision problems appear naturally in life, in general, and in engineering design in particular. However, the choice of criteria is not a trivial matter and is in fact an important part of the modelling process. If X is the vector of design variables defining a design, the multiple objective design problem can be stated as

$$\text{maximise} \quad F(X) = [f_1(X) f_2(X) \ldots f_k(X)]$$

$$\text{subject to} \quad X \in \Omega$$

$$\Omega = \left\{ X \left| \begin{array}{ll} g_i(X) \le 0 & i = 1,2 \;\; \ldots, m_1 \\ h_j(X) = 0 & j = 1,2 \;\; \ldots, m_2 \\ X = [x_1, x_2, \ldots x_n]^T & \end{array} \right. \right\} \qquad [1.1]$$

From the simple formulation above it is clear that just as the choice of X determines F(X) the choice of F(X) determines the most preferred value of X. In other words, the criteria have no objective existence by themselves except as a statement of the designer's view of the design situation. If the criteria change the design may be considered to be concentrating on a different set of performance aspects and this obviously has a significant bearing on what is perceived to be an efficient solution. These criteria are often arrived at by a method of hierarchical decomposition, starting with an overall objective like "efficiency" at the top and then describing this in terms of more meaningful measures of performance which in turn may be sub-divided into even more detailed aspects of performance. This is related to the advice given by Suh [Suh 1990] regarding the care needed to define appropriate functional requirements in design. Suh describes design in his axiomatic approach as a mapping between what we want to achieve and how we wish to achieve it – the former being represented by the *functional requirements* (FR) and the latter *by design parameters* (DP) as shown in Figure 1.3.

Suh's first axiom defining a good design stipulates independence of FRs. It is obvious that this is not always achievable, but the choice of FRs will definitely affect the design. In an alternative statement of this axiom an acceptable design is described as one in which the DPs and FRs are related in such a way that specific DPs can be adjusted to satisfy their corresponding FRs without affecting other functional requirements. Again this degree of independence may not be achievable but even if it were possible a specific functional requirement can be a complex concept requiring multiple measures of performance to represent it and choice of any set of these measures would affect the solution chosen.

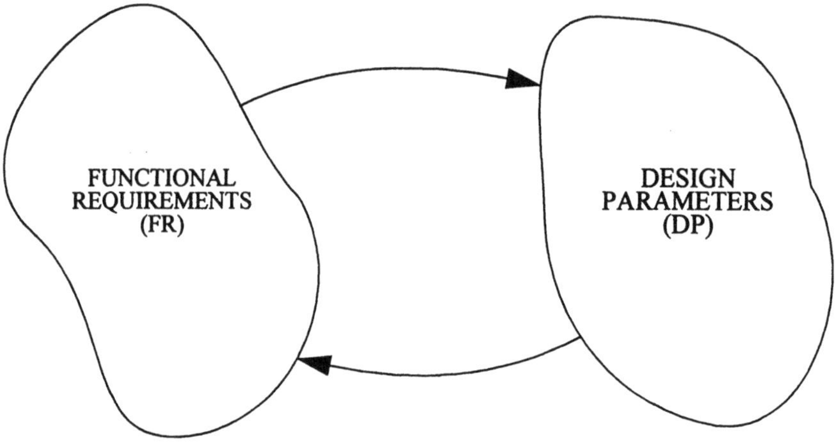

Figure 1.3 *Design as a mapping process*

Although the prevailing terminology in this general area of work is settling down to some sort of order there is as yet no universal agreement over all the descriptors used. Chapter 2 provides a simple guide to the relationship between various terms used within MCDM. In general, a criterion may be defined as a concept that allows comparison of alternatives according to a particular significance axis or point of view [Bouyssou, 1990].

Keeney and Raiffa [Keeney, 1976], amongst others, provide some desirable properties for attributes that may be used to compare alternative actions. For example, the attributes must be

complete:	so that all pertinent aspects of the decision problem are presented
operational:	in that they can be used in a meaningful manner
decomposable:	if disaggregation of attributes is required or is desirable
non-redundant:	so that no aspect of the decision problem is considered twice
minimal:	such that there is no other attribute set capable of representing the problem with a smaller number of elements.

It would be difficult at the best of times to conclusively show that all of the above properties are characteristic of the modelling strategy employed in any situation but it is important to note that the choice of alternative appropriate sets of criteria is an important design issue even though it may be difficult to anticipate the precise influence of a chosen set on the final solution. Given the intimate relationship between the sort of data available and the "decision rule" alluded to in Section 1.1 guiding the identification of the most preferable solution, the choice of the appropriate MCDM method also becomes of crucial importance. The aim of this book is not to present all the existing methods, as there are various texts that do this already, but to explore the principal issues that an engineering designer or decision maker needs to address in adopting different approaches. In pursuance of this aim certain methodological contributions by the authors and some existing methods are examined in the context of the overall structure of MCDM.

Chapter 2 describes the development of MCDM in greater detail but some of the key concepts in engineering design decision making have already emerged on the basis of discussions so far and these may be summarised as

- choice of criteria and choice of method
- need to capture the priority ordering of the designer
- clear and objective decision rules encapsulated in methods to produce results in a consistent manner.

1.2.3 Issues of Complexity, Subjectivity and Uncertainty

Some other complicating factors that appear in many decision making problems in engineering design are those related to the complexity of the task, the need to take account of subjective as well as objective factors and the inherent uncertainty in a given situation.

Complexity is most often handled in engineering by some form of decomposition. As already discussed in Section 1.2.2 above the development of appropriate criteria may be effected by hierarchical decomposition. In a parallel manner the design task itself can be hierarchically decomposed to an appropriate level of detail that is capable of being expressed by a set of independent attributes. Several of the approaches described later in the text makes use of this structure.

Design decision making often requires the examination of subjective as well as objective criteria [Table 1.1]. For example, the handling properties of a motor-cycle may be described in terms of linguistic descriptors rather than numerical values. The linguistic variables need to be considered in as objective a manner as possible, in that judgements made have to be converted into objective data without loss of the subjective feel of the judgements made. This is an important consideration as the whole domain of MCDM is largely about keeping the designer or DM in the loop and designers need to express intuitive judgements and have them evaluated along with more objective criteria.

Table 1.1 *Objective and Subjective Attributes for a Motorcycle Choice Problem*

<table>
<tr><td rowspan="2">types of attributes</td><td rowspan="2">definition of attributes</td><td colspan="2">unit factors</td></tr>
<tr><td>composite factors</td><td>basic factors</td></tr>
<tr><td rowspan="4">quantitative</td><td>price</td><td colspan="2">pounds</td></tr>
<tr><td>displacement</td><td colspan="2">cc</td></tr>
<tr><td>range</td><td colspan="2">miles</td></tr>
<tr><td>top speed</td><td colspan="2">mph</td></tr>
<tr><td rowspan="14">qualitative</td><td rowspan="5">engine</td><td colspan="2">responsiveness</td></tr>
<tr><td colspan="2">fuel economy</td></tr>
<tr><td colspan="2">quietness</td></tr>
<tr><td colspan="2">vibration</td></tr>
<tr><td colspan="2">starting</td></tr>
<tr><td rowspan="9">operation</td><td rowspan="4">handling</td><td>steering</td></tr>
<tr><td>bumpy bends</td></tr>
<tr><td>manoeuvrability</td></tr>
<tr><td>top speed ability</td></tr>
<tr><td rowspan="2">transmission</td><td>clutch operation</td></tr>
<tr><td>gearbox operation</td></tr>
<tr><td rowspan="3">brakes</td><td>stopping gear</td></tr>
<tr><td>braking stability</td></tr>
<tr><td>feel at control</td></tr>
</table>

Assessment of such subjective measures of performance necessarily involve a degree of uncertainty in that it is never absolutely clear whether an "excellent" rating is very different from a "very good" one. It is thus necessary to address these decision situations by a method of analysis that maintains the essentially subjective assessments but combines such assessments with objective analysis in an analytically rigorous and consistent manner. A common method of dealing with such subjective data is to convert it to objective values using a numerical scale where "excellent" stands for 10, say, and "unacceptable" stands for 0. But this conversion is essentially opposed to the subjective nature of the evaluations. A more satisfactory way of dealing with this situation would be to use a number scale but interpret the grade "excellent" to denote a range between 10 and 8, say, with different degrees of belief associated with the various grades or their associated numerical values within that range. This type of approach is outlined in somewhat greater detail in Chapter 3.

Just as complexity in objective analysis is dealt with using hierarchical decomposition, subjective evaluations can just as conveniently be handled by decomposing a complex qualitative attribute into lower level sub-attributes to make assessments more meaningful and manageable for the designer or DM. Thus the assessment of the *engine* in Table 1.1 can be on the basis of sub-attributes like *fuel economy* or *quietness*.

1.3 Design Selection vs Design Synthesis

In the mathematical statement of the multiple objective design problem in Section 1.2.2 the vector X is treated as the vector of design variables whose values have to be found in an effort to maximise a range of objectives. Whether the set of X values from which a selection is to be made is finite and known, or implicitly defined by a set of requirements describes the two broad classes of design decision problems – selection and synthesis. However it is easy to see that the type types of problems are closely related.

Let us examine the problem of selection with reference to a hypothetical situation represented by Table 1.2, where $a_i, i = 1,2,3,4,5$ are different alternatives and $y_j, j = 1,2,3$ are the various attributes of the five alternatives. For the sake of simplicity all attributes are considered to be maximising.

Table 1.2 *A General Selection Problem*

Alternatives	Attributes		
	y_1	y_2	y_3
a_1	10	60	0.4
a_2	4	120	0.7
a_3	6	30	0.9
a_4	3	50	0.6
a_5	6	90	0.5

It is obvious that alternative a_4 is dominated by alternative a_2 in that all the attribute values of a_2 are greater than the corresponding attributes for alternative a_4. It is clearly illogical to prefer a dominated solution over a non-dominated one. So if a_4 is eliminated from the list the remaining solutions constitute a non-dominated set. There may be other possible solutions that dominate the remaining alternatives a_1, a_2, a_3 or a_5 and if these are known it would be sensible to include them and drop whichever of the existing alternatives that happens to be dominated. However, if no such dominating solution exists or needs to be considered for the problem in hand the remaining solutions a_1, a_2, a_3 and a_5 would constitute an efficient set in that neither of any pair of alternatives have better values of one or more attributes without simultaneously having poorer values for other attributes.

The above problem can be viewed in another way. If instead of having a set of feasible efficient alternatives we had a mathematical model for generating solutions then a_1, a_2, a_3 and a_5 would have been precisely the type of solutions generated by maximising a linear combination of the objectives

$y_j = f_j(X), j = 1,2,....,k$ of the form $\sum_j w_j f_j(X)$ subject to $X \in \Omega$ if the feasible region defined by Ω was convex [See 1.1].

In this formulation, if $\sum_j w_j = 1$ then for $w_j = 1$ for j = 1, 2 or 3 we would obtain solutions a_1, a_2 and a_3 respectively as these are solutions with maximum values of f_1, f_2 and f_3 . Solution a_5 is obtained when $w_j \neq 1$ for any *j*.

There is thus a symmetry of argument in the formulations for selection and synthesis and there is thus no absolute necessity to distinguish between the two approaches. However, the decision problems in design selection and design synthesis are qualitatively distinct and there is merit in looking at them separately while keeping in mind the essential synergy of ideas.

1.4 Outline of the Book

The rest of the book is organised along the following lines.

In Chapter 2 the main methodological issues related to Multiple Criteria Decision Making are briefly examined. This provides the backdrop for the detailed examination of the individual methods and their application in design. It may be observed, broadly speaking, that the systems engineer or DM is largely interested in finding guidelines for combining judgements with formal procedures of decision making and the behavioural scientist is interested in the psychology of judgement when multiple criteria trade-off is involved. Since the design engineer is largely interested in integration activities it is necessary for him to be familiar with the principal issues in both areas.

In Chapter 3 some selection or Multiple Attribute Decision Making (MADM) techniques are examined in detail. Since a majority of decisions in real life involve selection of some kind it is important to examine how alternative methods lead to different insights and make different demands in terms of availability of data.

Chapter 4 deals with design synthesis or Multiple Objective Decision Making problems where alternative priorities over objectives lead to different design solutions. Design synthesis is commonly an interactive task in which the designer might, by examining the available range of options, revise his sense of priorities, and this in turn, of course, affects the range of available solutions. Articulation of preferences in MCDM can be either *a priori* or *progressive* or *a posteriori* but in design synthesis the interactive nature of the decision making process often makes progressive articulation even more relevant.

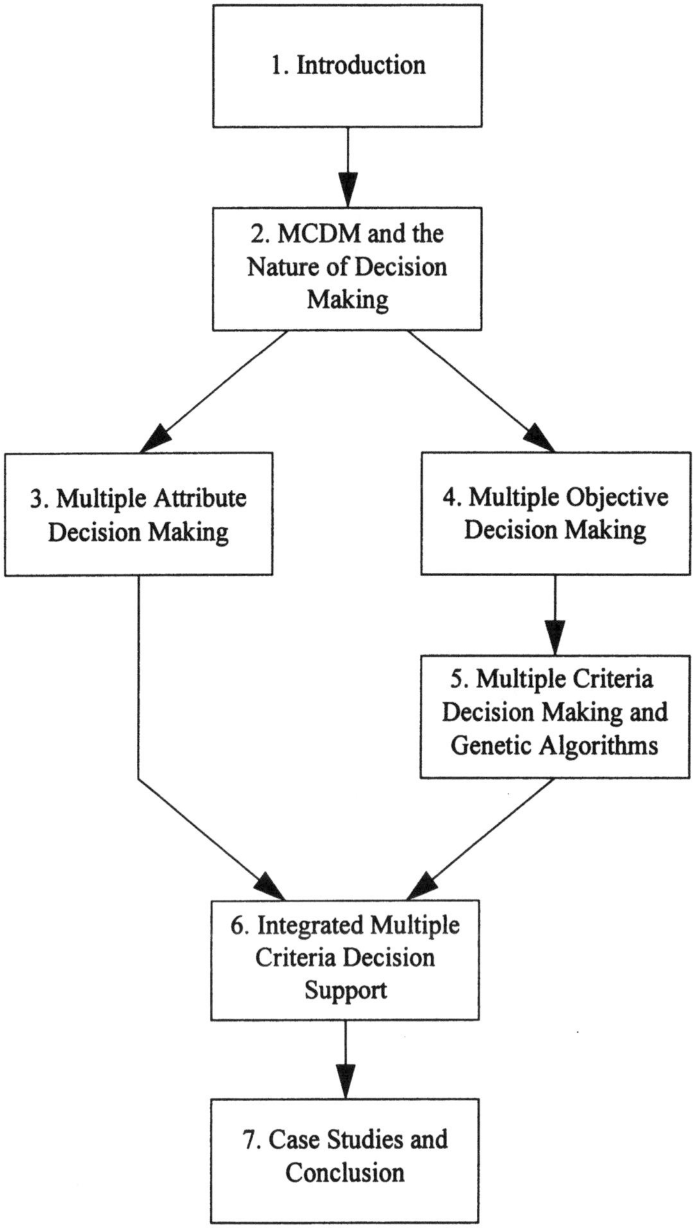

Figure 1.4 *Book Structure*

Having established the methodological base in the earlier chapters, Chapter 5 relates to the use of emerging adaptive optimisation tools that are particularly helpful when multimodal, discontinuous and non-mathematical models may be involved. The development and use of Multiple Criteria Genetic Algorithms is used as the example vehicle for critically examining some practical issues of application.

Chapter 6 deals with the principal features of a computer based decision support system (DSS) built on the above methodological base. The DSS provides guidance on appropriate solution strategies and provides some ideas on future areas of development. Some practical aspects of using decision support systems are highlighted.

In Chapter 7 the multiple criteria paradigm is examined critically from the engineering designer's point of view through a couple of case studies to demonstrate how MCDM can be helpfully imbedded within a broad design scenario. Every text aims to provide some answers but in the process of doing so it poses new questions. Some of these related issues are briefly examined in this chapter .

Figure 1.4 shows the general structure of the book.

There are various ways of using this text, but the individual reader should follow it linearly, if possible, leaving out Chapter 6 in the first instance. A one semester course would also use the text linearly but elect to concentrate on a set of methodologies from Chapters 3, 4 and 5 to suit the needs of the course.

2
MCDM and the Nature of Decision Making in Design

2.1 Introduction

In Chapter 1 a brief outline is provided of the motivation behind the use of multiple criteria decision making (MCDM) techniques. Decision making, in general, and in engineering design, in particular, can be helpfully visualised as a collection of activities that relate to choice in the context of competing technical or functional requirements. The options may either be available and finite in number, as in consulting a catalogue, or they may need to be synthesised, as in engineering design. In any event, the implicit assumption that is often made is that the requirements in question are mutually compatible. This is the domain of classical optimisation in that it is being taken for granted that the stated requirements are mutually compatible or can be made so, although even in classical optimisation there is an implicit acknowledgement of conflict in only being able to design for a stated scenario.

Consider, for example, the issue of deciding on ship size for a particular set of operating conditions using, let us say, a cost-based criterion. Figure 2.1 shows how optimal size is dependent on the operating scenario in question. An operating scenario can be thought of as a combination of market share, fuel costs, other operating costs and port conditions that defines a particular identifiable operating environment.

If operating condition OP1 prevails and one is confident that this condition will not change over the lifetime of the vessel in question, it would obviously make sense to choose Solution 1 as Figure 2.1(a) shows. Departure from the optimal solution in this example is assumed to lead to parabolic penalties in terms of cost of transportation. If it was equally likely that either operating condition OP1 or OP2 would come about then the sensible solution to choose would not be either Solution 1 or Solution 2 but Solution 4, so that under either scenario, one would not be doing too badly. Choosing either Solution 1 or 2 would lead to large cost penalties if the scenario for which the ship has been optimised did not transpire. These arguments can obviously be generalised to a range of operating conditions.

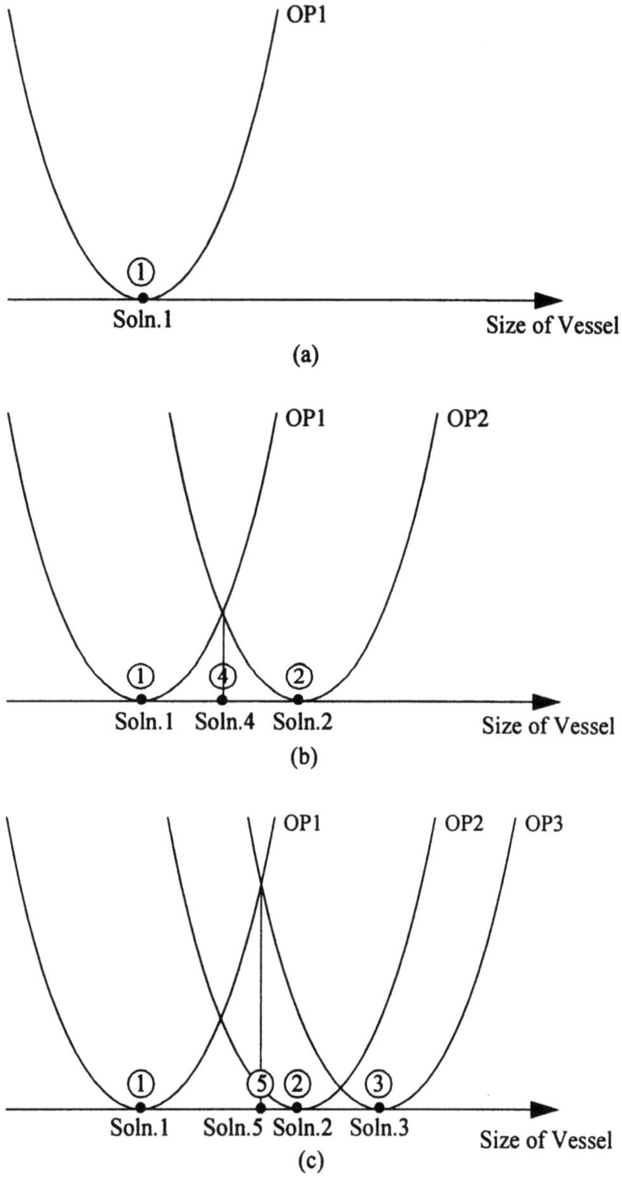

Figure 2.1 *Optimal ship size for changing operating conditions*

Thus Figure 2.1(c) shows the optimal solution for OP_3 and Solution 5 is the compromise Solution if OP_1, OP_2 or OP_3 were equally likely. Solutions 4 and 5 represent solutions of minimum regret in that no matter what happens one is doing as less badly as possible. The above set of considerations, with minor modifications relating to domain of application, is true for all areas of decision making, particularly in engineering, because an artefact designed and built to a specific set of market expectations often find themselves having to cope with a completely different set of conditions. Designers and decision makers (DMs) sometimes cope with this by building in margins. But whatever the remedy that may be tried it is obvious that what is being coped with is the influence of alternative operating scenarios. Thus even in the world of classical mono-criterion optimisation, different operating scenarios can lead to conflicting requirements, and these have to be dealt with.

2.2 Pareto Optimality: What are the Options?

On the basis of the introductory observations above it can be said that it is necessary, in all except the most trivial applications, to consider how systems perform in a range of plausible conditions. This is why most of the sensitivity studies that accompany optimisation are done. For most real applications, finding the optimal solution is only part of the exercise as it is necessary in most cases to find how sensitive a solution is to changing conditions. Finding a robust optimum whose performance is good and also relatively insensitive to changing conditions is a very important concern for engineering designers.

From that observation it is a short route to the assertion that requirements in design of any kind are often potentially in conflict. This is because there are few, if any systems that can combine the best of all performance aspects for all possible scenarios in the same design. If such utopian solutions exist then the obvious answer would be to go for them. But life being the way it is, good values of some criteria inevitably go with poor values of others. The aim in multiple criteria decision making is then to find the best compromise solution. The process of compromise must reconcile the potentially conflicting requirements in the light of stated or implicit priorities of the DM. Even in those situations where potentially conflicting requirements are harmonised by setting clearly achievable threshold values one is faced with the task of choosing the best solution that meet all the requirements. This process of selection is just a limited version of the problem of compromise in that the final choice has to take account of the differential values to the DM of the various attributes of the candidate acceptable options.

This process of compromise or conflict resolution in decision making can be conveniently visualised by examining a simple problem in terms of two criteria, as shown in Figure 2.2.

Without loss of generality all criteria in multiple criteria decision making can be thought of as maximising, as is implicit in Figure 2.2, given that it is easy to convert a minimising criterion to a maximising one by changing the sign of the criterion in question In Figure 2.2 it is clear that if each criterion is maximised in turn the solutions obtained would be A and B respectively. The ideal solution, combining the performance of these two solutions would be I. In general, as observed already, this solution is nearly always unattainable, because physical and modelling constraints do not allow this to happen. One is therefore reduced to the task of finding the "best" solution within the feasible region, shown shaded in Figure 2.2. This is a region that is defined by the functional constraints.

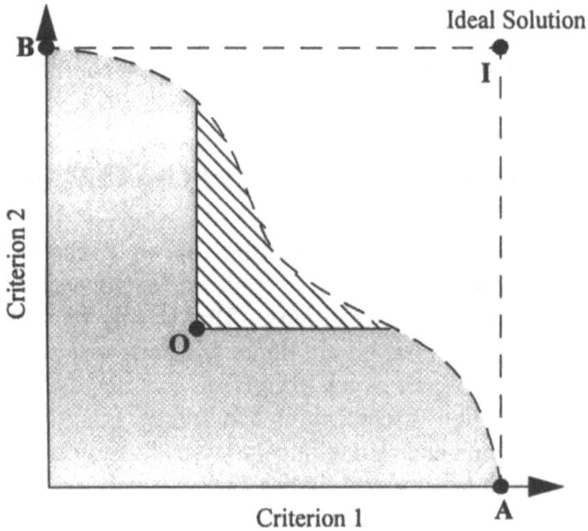

Figure 2.2 *The Pareto Front*

It is clear, however, on the basis of commonsense arguments, that all solutions in the hatched wedge around the solution represented by point 0 will be feasible and superior to it. This is because all solutions within the hatched wedge are better than solution 0 at least with respect to one criterion if not in terms of both. Solution 0 may therefore be considered to be dominated by all solutions within the hatched wedge. As point 0 moves towards the boundary separating the infeasible from the feasible solutions, however, the set of dominating solutions represented by the area of the shaded wedge reduces. Thus when point 0 lies on the boundary there is no solution that can be said to dominate it. This is true of every point that lies on the boundary. Points on the boundary therefore represent non-dominated solutions where improvement in terms of one criterion can only be at the expense of the other. This boundary is referred to as the Pareto front and contains all solutions of interest because no point anywhere except on this boundary can be anything other than either dominated or infeasible.

The solution strategy for multiple criteria decision making is then to define this front and then obtain the "best" point on it. If the ideal decision is defined in some way, MCDM is basically about coming as close as possible to the ideal while remaining within the feasible region. In doing so the principal options are:

(i) Obtain solutions on the Pareto front by multiple criteria search and then use a multiple criteria selection strategy to find the "best" solution on the basis of priority structure of the DM. This is tantamount to finding out the nature of the solution space before selecting a solution.

(ii) Express preference structure directly, indirectly, implicitly or interactively to convert a multiple criteria problem into a more manageable form and then find the "best" solution by appropriate synthesis models. This is basically about making up one's mind about what one likes and finding a solution in harmony with those preferences.

There is a large array of methods that help this process of selection and synthesis, and Chapters 3, 4 and 5 deal with some of them in greater detail.

In doing the above, however, several issues have to be taken into account. It has to be borne in mind that MCDM requires that

- both selection and synthesis be considered.
- formal methods be used for priority assignment or preference communication.
- problems of technical trade-off in the multiple criteria domain be facilitated.

Some general observations are necessary at this stage. Users of decision making tools of any kind are largely driven by at least two principal considerations. Firstly, there is the need to process the data relevant to the problem in hand in such a way that as much information is extracted as possible to assist decision making. This is what the various MCDM methods do in effect. This activity of information extraction naturally involves the processing of the relevant data that defines the problem. This gives rise to the second of the two considerations.

Processing of the data inevitably results in a distance between the data and the DM, and the more elaborate the information processing and the more subtle the decision making process the less feel for the problem the DM has in practice. There is thus a tension here between the need to be making as much use as possible of the available information while keeping the whole process as "hands on" as possible. There is not much point in establishing a very involved decision procedure that puts so much distance between the problem and the DM that the resulting decisions are poorer than they would otherwise be using a simpler and more transparent procedure.

Over and above these two overarching requirements there is the need for repeatability of results. In other words the same questions asked and answered in the same way should lead to the same results. In methodological terms this is tantamount to saying that the same methods used in the same way should repeatedly produce the same decisions. It is quite obvious that a transparent procedure that does the job simply but effectively in response to the two principal requirements identified above should also produce the repeatability that is the hallmark of all methods that truly assist decision making in the technical or other domains.

2.3 MCDM Methods and Some Key Terminology

In pursuing the above it should be borne in mind that there is no universally acceptable mode of representation and communication of the preference information referred to above because each decision situation has its own special features and thus requires information of a different kind. This gives rise to a large number of analytical techniques and this is what the rest of the book is all about. In general terms, however, it is clear that decision making in the presence of multiple, potentially conflicting criteria can be broadly classified as one of two types:

- selection of an alternative from a menu or catalogue based on prioritised attributes of the alternatives (i.e. multiple attribute decision making or MADM)

- synthesis of an alternative or alternatives on the basis of prioritised objectives (multiple objective decision making or MODM)

To clarify the terminology the following diagram may be helpful.

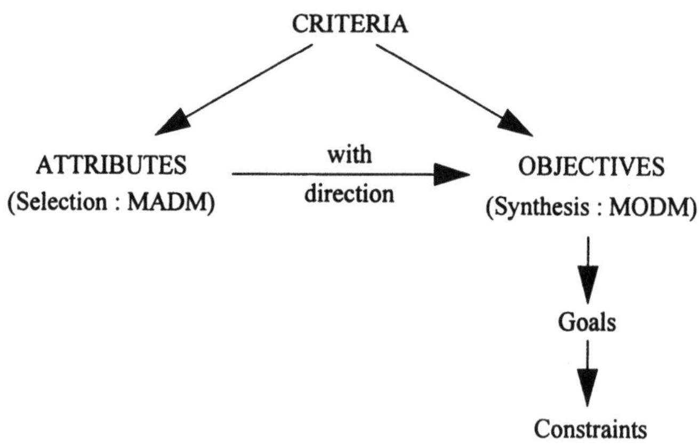

Figure 2.3 *Multiple Criteria Decision Making*

A criterion can be thought of as any measure of performance for an alternative. When there is a set of alternatives to choose from, the choice is most conveniently made by using some form of static or moving weights to represent the contribution of the various common attributes of the alternatives.

Selection problems can thus be thought of as multiple attribute decision making (MADM) problems. Most readers will be familiar with the common scheme of comparing alternatives on the basis of weighted sum of normalised attributes. The normalising of the attributes ensures their comparability, as otherwise high numbered attributes would make disproportionate contributions to the overall score. Utility functions may be visualised as moving weights so that the relative contributions made by different attributes to the ranking of alternatives change with the attribute values themselves.

When there is no list of solutions to choose from but only a list of requirements to meet, it is appropriate to think in terms of objectives. An attribute with direction is an objective. Thus cost and weight are attributes but the aim of minimising cost and minimising weight are objectives. As problems of synthesis are largely about meeting objectives, prioritised according to the relative importance of the objectives set by the DM, design or synthesis problems can be thought of as multiple objective decision making (MODM) problems.

All decision problems can be classified as belonging to one of these two broad classes.

Pursuing the terminology a bit further it can be asserted that if the thresholds of the objectives are flexible in the sense that the requirements represent aspirations (e.g. some non-statutory requirement relating to some desirable but non-crucial aspect of performance) rather than hard bounds then the decision problem reduces to a format that is most conveniently handled by techniques like goal programming in which multiple objectives are addressed by minimising the weighted sum of deviations from stated goals or threshold values of performance. In fact, if some of the bounds are hard (e.g. permissible stress) and some are flexible, it is a situation that is well suited to generalised goal programming. If, on the other hand, the thresholds represent strict bounds only, the objectives then become constraints and the formulation is the domain of classical optimisation where conflicting requirements can only be handled by negotiating the bounds or thresholds.

The above arguments demonstrate how the MCDM approach to decision making, in general, and in engineering design, in particular, is a generalised approach that accommodates classical optimisation but transcends its limitations. The MCDM approach, by addressing the DM's priorities, makes the underlying trade-offs between criteria transparent and capable of convenient manipulation, and this can often lead to better decisions overall.

Before launching into the mechanics of MCDM methodology and applications it is sometimes helpful to remember that just as in the mono-criterion paradigm the objective function value is a function of the design variables, so also in the multiple criteria domain the overall attractiveness of a solution or alternative is a function of the individual criteria or aspects of performance. In some respects, MCDM is about shifting from variable-to-objective functional relationships to relationships relating the overall score of an alternative to the contributing criteria. In practice, though, MCDM is the more general and helpful construct as it deals with trade-off in all its aspects. This allows the DM or engineering designer to explore the consequences of applying alternative performance models in terms of the design or decision variables while also allowing alternative specifications to shape the emerging solutions.

2.4 Concluding Comments

Having examined both the background to problem solving using MCDM and the manner in which it may be used in design decision making, the scene is now set for examining how the individual methods work in practice. This is examined in the next three chapters. The methods are presented in a compact but reasonably rigorous manner and demonstrated on applications. This is to address the basic requirements of the task of examining such methods within the context of real life decision making.

3
Multiple Attribute Decision Making

3.1 Problem Formulations and Method Classification

3.1.1 MADM Problems

A multiple attribute decision making (simply MADM) problem usually comprises a finite number of explicitly given alternative designs and a set of performance attributes. Design selection involves either choosing the most favourable design from the alternative set or ranking all the alternative designs with regard to all attributes. A MADM problem may have either qualitative or quantitative data. More generally, MADM problems may involve both types of data, and approaches for dealing with these will be investigated in Section 3.4.

The methods of communication of the attribute values and their relative importance determine the broad classification of these problem types. If in a MADM problem each alternative is not numerically measured over every attribute but relatively evaluated by means of subjective judgements, it is regarded as a qualitative evaluation problem. One example of such qualitative MADM problems is where each pair of alternatives are compared with regard to every attribute based on certain standard for comparison.

Figure 3.1 shows an ordinary MADM problem with n alternatives (a_i, i=1, $\cdots$,n) and k attributes (y_j, j=1, $\cdots$,k). Each pair of alternatives (a_i, a_l, i, l=1, $\cdots$,n; $i \neq l$) are compared with respect to every attribute (y_j, j=1, $\cdots$,k). If m_{il} represents the relative importance of a_i over a_l with respect to y_j, a pairwise comparison matrix for all the n alternatives in terms of the attribute y_j can be formulated as in equation (3.1). The MADM problem is then represented by k pairwise comparison matrices for the k attributes.

$$M = \{m_{il}\}_{n \times n} = \begin{bmatrix} 1 & m_{12} & \cdots & m_{1n} \\ m_{21} & 1 & \cdots & m_{2n} \\ \cdots & \cdots & \cdots & \cdots \\ m_{n1} & m_{n2} & \cdots & 1 \end{bmatrix} \tag{3.1}$$

where $m_{lh} = 1/m_{hl}$ for all l, h=1, $\cdots$,n;, due to symmetry of comparison.

Figure 3.2 shows a more general hierarchical MADM problem with a multi-level attribute structure, multiple decision makers and incomplete pairwise comparisons which imply that not all of the lower level attributes (or alternatives) are related to each of their immediate upper level attributes. Numerically the problem can still be represented by the set of pairwise comparison matrices for all the lower level attributes (or alternatives) with respect to each of the upper level attributes.

Let y_j be the *jth* attribute (j=1, $\cdots$,k) and a_i the *ith* alternative design (i=1, $\cdots$,n). Suppose y_{ij} stands for the value of an attribute y_j with respect to a design a_i (i=1, $\cdots$,n; j=1, $\cdots$,k). Then a quantitative MADM problem of ranking n alternative designs based on k attributes may be represented using the following decision matrix, as shown in Table 3.1.

Given the relative importance of the attributes, the available alternatives can be ranked by a variety of methods as discussed below.

3.1.2 Classification of MADM Methods

Classification of MCDM techniques provides a framework on which advice can be based as to the aptness of a method for a specific problem situation. For generic classification of MADM techniques, several schemata have been proposed. For instance, Teghem et al. used six criteria to categorize MADM methods [Teghem et al. 1989]. Ozernoy used a hierarchical model for selecting the most appropriate MADM techniques [Ozernoy 1987].

However, there has been no single, widely adopted classification of MADM methods. This section is not intended to suggest a generic classification schema either but to propose a specific classification framework for application of MADM methods in engineering design. Such a framework is shown in Figure 3.3. The methods listed in Figure 3.3 are only some of the methods which have been developed in the past two decades. Within the context of the classification system, some rules for selecting MADM methods may be discussed as follows.

The classification of MADM methods, as shown in Figure 3.3, is mainly based on what input evaluation data is required and how the designer's preferences are acquired and represented. It has already been noted that the diversity of evaluation data and preference representation can be used to distinguish one MADM method from another. This is because the characteristics of a MADM method in terms of preference representation and data requirement decide to a

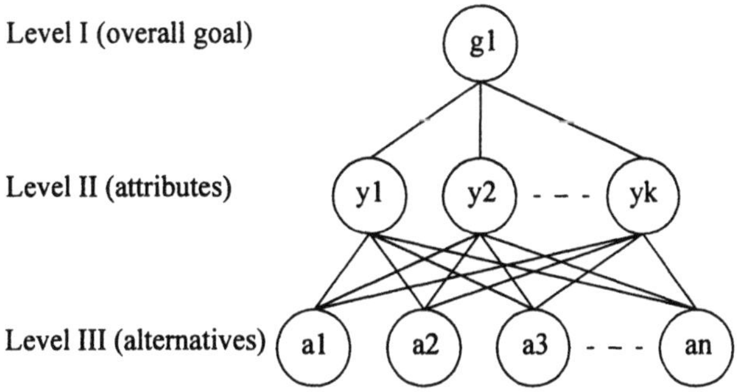

Figure 3.1. *A hierarchy with single-layer attribute structure and complete comparisons*

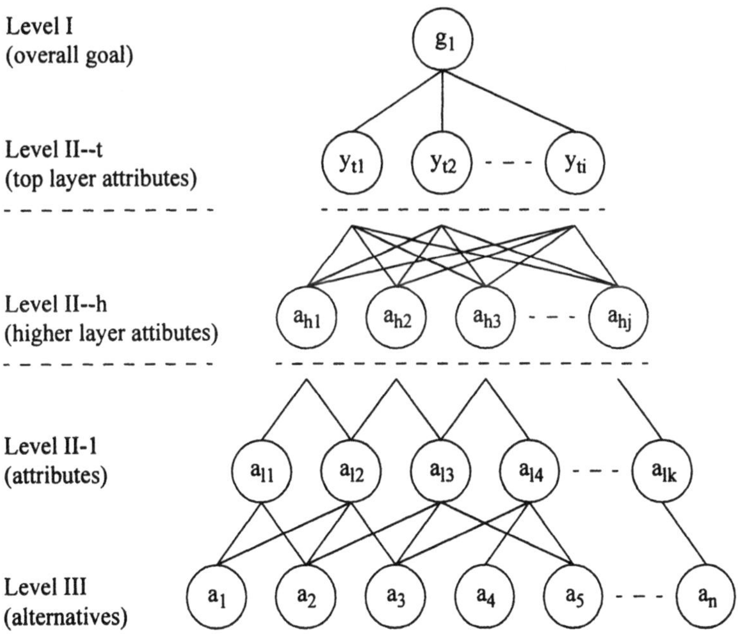

Figure 3.2. *A hierarchy with multi-layer attribute structure and incomplete comparisons*

Table 3.1 *Decision Matrix*

Alternative designs	Attributes			
	y_1	y_2	...	y_k
a_1	y_{11}	y_{12}	...	y_{1k}
a_2	y_{21}	y_{22}	...	y_{2k}
...	...	...	...	...
a_n	y_{n1}	y_{n2}	...	y_{nk}

large extent the manner of interaction between the method in question and a designer. It is probably true to say that for a designer the mathematics and computational steps involved in a MADM method are less important than the interaction procedure.

The rules for selecting an appropriate MADM method can therefore be divided into two subsets. One subset of rules can be used to differentiate the ways in which preference information is elicited and represented in a MADM method. The other can be used to distinguish the types of input evaluation data which can be processed in a MADM method. Given the same data type, methods may still differ in terms of data processing strategies or decision rules.

In some MADM methods no preference information is needed and only certain simple decision rules are employed to rank or to sort alternative designs. In some other methods preference information is required and represented in different ways. Relative weights and utility functions are widely used for representing preferences. Standard levels of attributes provide another means for representing preferences. In some methods it is required that weights be given beforehand. A few techniques are available for assigning appropriate weights.

Input evaluation data may be either quantitative or qualitative. Qualitative data includes attributewise rankings of alternatives and pairwise comparisons between alternatives or between attributes. Quantitative data can be represented by a decision matrix in which every alternative is evaluated numerically with respect to each attribute.

Figure 3.4 illustrates some of the rules of choice for selecting an appropriate MADM method based on features like the acquisition and representation of preference information and requirement of input evaluation data. For instance, a choice rule for selecting the UTA method may be listed as follows

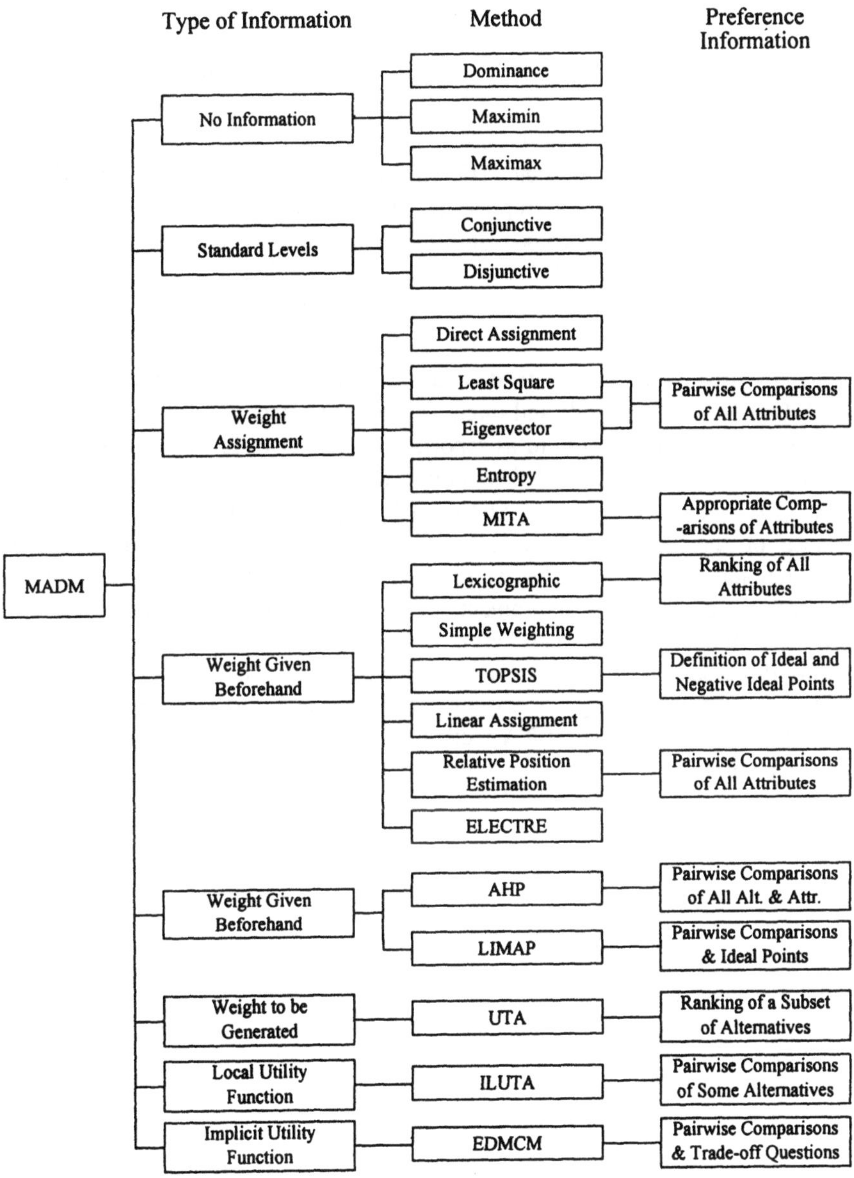

Figure 3.3 *Classification of MADM Methods*

if *preferences can be elicited in terms of the subjective ranking of a subset of alternatives, and*

if *preferences can be represented by additive utility functions,*

then *the UTA method is suggested.*

Some methods may require the same type of preference information and input evaluation data but use different decision rules. In this case, the decision rule employed by a method will be used to distinguish the method from other methods. For instance, both the TOPSIS method and the ELECTRE method require a decision matrix to represent input evaluation data and use relative weights to represent preference information. However TOPSIS defines the relative closeness to an ideal design as the decision rule for ranking alternatives while ELECTRE uses concordance and discordance indices.

Figure 3.3 lists 16 MADM methods and five weight assignment techniques from which four MADM methods and three weight assignment techniques are selected for the decision support system as reported in Chapter 6. It would be of interest to examine the reasons behind this choice. The TOPSIS method and the revised ELECTRE method (christened CODASID [Yang, Sen et al. 1997]) are selected because of their simple logic, full utilization of information and systematic computational procedures. The weights required by the two methods can be obtained using the direct assignment technique, the eigenvector technique, the entropy technique, or the new minimal information method [Sen and Yang 1994a] discussed later in this chapter. The selected AHP method provides a simple and practical way to acquire, represent and analyze input data and preference information. The UTA method is also chosen as it adopts a different way of eliciting and representing preference information, which may suit some designers and design decision making situations.

3.2 Techniques for Weight Assignment

In the decision support system, three weight assignment techniques have been adopted and they are the direct assignment technique, the eigenvector method and the entropy method. A new technique has also been developed which only uses minimal amount of preference information. In Subsection 3.2.1 below these weight assignment techniques are discussed.

3.2.1 Direct Assignment

An experienced decision maker may be able to directly evaluate the relative importance of one attribute over others using certain evaluation standard. For instance, a 10-point scale may be chosen and calibrated in one of several ways,

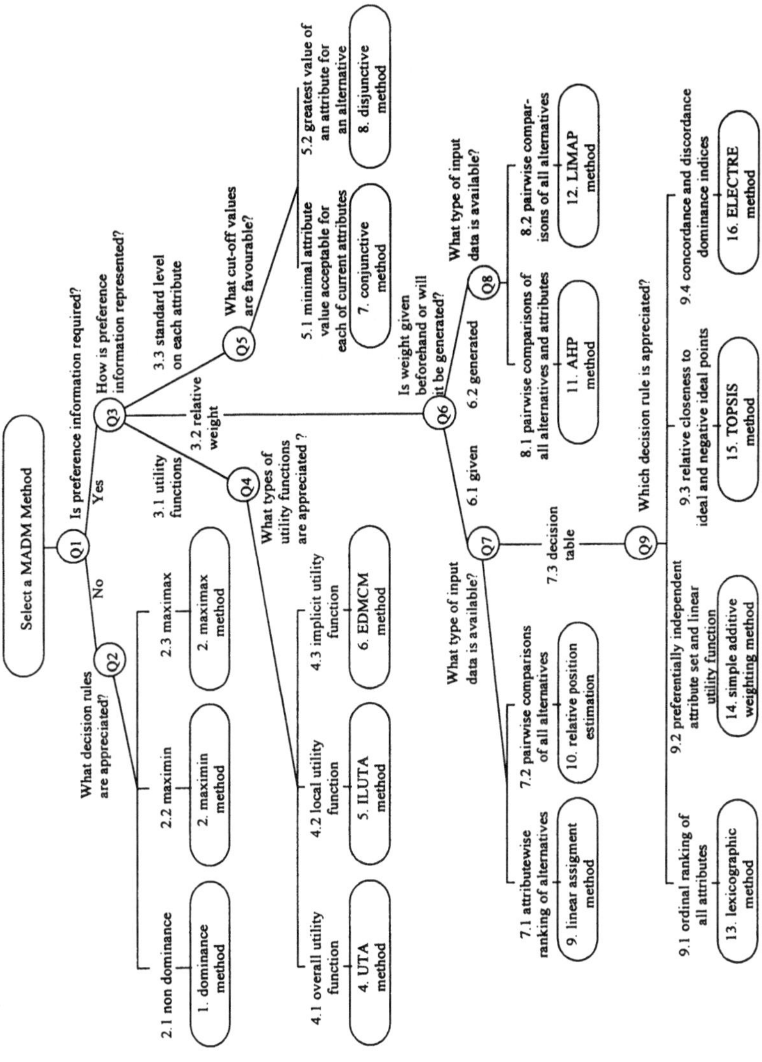

Figure 3.4. *Decision Tree for Selecting an MADM Method*

an example of which is shown in Figure 3.5 [Hwang and Yoon 1981]. It should be noted that the numerical assignment given in Figure 3.5 is arbitrary. Many other scales are possible. Besides, this type of scaling assumes that a scale value of 9.0 is three times as favourable as a scale value of 3.0. It also assumes that the difference between "unimportant" and "important" is the same as the difference between "average" and "very important". None of these assumptions, of course, need be true in a given decision situation.

Attribute evaluation	Value
extremely unimportant	0
very unimportant	1.0
unimportant	3.0
average	5.0
important	7.0
very important	9.0
extremely important	10.0

Figure 3.5 *Assignment of Values* for *A 10–Point Scale*

To demonstrate how weights may be assigned directly, take for example a fighter aircraft selection problem. Suppose six attributes are taken into consideration in the problem and they are "*Maximum speed* (f_1)", "*Ferry range* (f_2)", "*Maximum pay load* (f_3)", "*Acquisition cost* (f_4)", "*Reliability* (f_5)" and "*Maneuverability* (f_6)". The relative importance of these attributes may be directly assigned by the decision maker on the basis of the scale defined by Figure 3.5. For instance, "*Maximum speed*" may be "very important", "*Ferry range*" and "*Maximum pay load*" may be between "very unimportant" and "unimportant", "*Acquisition cost*" and "*Reliability*" could be "unimportant" and "*Maneuverability*" could be "very important". From Figure 3.5, we may then have the following weights

$$W = [w_1 \;\; w_2 \;\; w_3 \;\; w_4 \;\; w_5 \;\; w_6]^T = [9 \;\; 2 \;\; 2 \;\; 3 \;\; 3 \;\; 9]^T \tag{3.2}$$

where w_i is the weight of f_i.

3.2.2 Eigenvector Method

The eigenvector method provides another way of eliciting and representing preference information. It is simple and easy to understand. To capture the DM's preferences, this method uses pairwise comparisons between attributes, represented by a comparison matrix as shown by equation (3.1). m_{lh} in equation (3.1) implies that attribute f_l is m_{lh} times as important as attribute f_h. m_{lh} for all $l, h = 1, \cdots, n$; $l \neq h$ need to be given by the DM. Relative weights w_i may then be obtained as the following normalized eigenvector

$$MW = \lambda_{\max} W \tag{3.3}$$

where $W=[w_1 \cdots w_k]^T$ and $\lambda_{\max}$ is the maximum eigenvalue of the comparison matrix M. The normalized eigenvector obtained by solving equation (3.3) may also be referred to as priority vector.

All pairwise comparisons in M will be consistent, or M will have complete consistency, if the following conditions are satisfied

$$m_{lh} = \frac{1}{m_{hl}} \quad \text{and} \quad m_{lk} m_{kh} = m_{lh}$$
$$\text{for } all \ l, h, k=1, \cdots, n; \ l \neq h \tag{3.4}$$

In this case, $\lambda_{\max} = n$. However, pairwise comparisons are normally inconsistent as the second part of formulae (3.4), or $m_{lk} m_{kh} = m_{lh}$, can be rarely satisfied for a problem of any reasonable size.

As such inconsistency may affect the accuracy of W, a consistency index is thus defined to measure the degree of inconsistency. The following algorithm provides a simple procedure to approximately calculate the normalized right eigenvector W with regards to the largest eigenvalue of the comparison matrix M as defined by equation (3.3).

Step 1: Provide an initially normalized vector $W^0 = [1 \ \ 0 \cdots 0]^T$ and let $t = 0$.

Step 2: Calculate a new eigenvector as follows

$$W^{t+1} = MW^t \tag{3.5}$$

Step 3: Calculate the maximum eigenvalue by

$$\lambda_{\max} = \sum_{i=1}^{n} w_i^{t+1} \tag{3.6}$$

Step 4: Normalize and update the eigenvector as follows

$$\overline{w}_i^{t+1} = \frac{w_i^{t+1}}{\lambda_{\max}}, \qquad w_i^{t+1} = \overline{w}_i^{t+1} \ \text{ for } all \ i=1, \cdots, n \tag{3.7}$$

Step 5: Calculate the error between the old and new eigenvectors and then check if

$$| w_i^{t+1} - w_i^t | \leq \delta \quad \text{for } all \ i=1,\cdots,n \tag{3.8}$$

where δ is a small non-negative real number (say δ=1.0×10^{-6}). If this is the case, go to step 6. Otherwise, let $t=t+1$ and go to step 2

Step 6: Calculate consistency index (CI) as follows

$$CI = \frac{\lambda_{max} - n}{n - 1} \leq 0.1 \tag{3.9}$$

as suggested by Saaty [Saaty 1988].

The above algorithm can be demonstrated by using the same example as examined in the previous section. From equation (3.2), it can be concluded that attribute f_1 is 4.5 times as important as attributes f_2 or f_3, 3 times as important as attributes f_4 or f_5, and equally important as attribute f_6. Other attributes can be compared similarly. The pairwise comparison matrix equivalent to equation (3.2) can thus be written as follows

$$M = \begin{bmatrix} 1 & 4.5 & 4.5 & 3 & 3 & 1 \\ & 1 & 1 & 0.667 & 0.667 & 0.222 \\ & & 1 & 0.667 & 0.667 & 0.222 \\ & & & 1 & 1 & 0.333 \\ & & & & 1 & 0.333 \\ & & & & & 1 \end{bmatrix}$$

Following the above steps, we can obtain the relative weights of the six attributes as follows

$$W = [0.3215 \quad 0.0714 \quad 0.0714 \quad 0.1071 \quad 0.1071 \quad 0.3215]^T$$

The corresponding largest eigenvalue is 6, or $\lambda_{max} = 6$. From equation (3.9), we have $CI = 0$. The results show that the above comparison matrix has complete consistency. This is because the matrix is constructed with reference to equation (3.2). In fact, the above weights are the same as those obtained by normalizing the weights given by equation (3.2) with the sum of all weights equal to one.

3.2.3 Entropy Method

If the input evaluation data of a MADM problem is represented by a decision matrix, as shown by Table 3.1, the entropy method may be used for assigning weights. The entropy idea is particularly useful for investigating contrasts in discrimination between sets of data [Hwang and Yoon 1981]. Since an outcome y_{ij} includes certain information content, the information content of the normalized outcomes of attribute j can be measured by means of an entropy value.

Suppose a decision matrix is defined by Table 3.1. Then, the normalized outcome of attribute j with respect to alternative i, p_{ij}, can be defined as follows

$$p_{ij} = \frac{y_{ij}}{\sum_{i=1}^{n} y_{ij}} \quad \text{for } \mathit{all} \ i, j \tag{3.10}$$

The entropy E_j of the set of normalized outcomes of attribute j is given by

$$E_j = -\alpha \sum_{i=1}^{n} p_{ij} \ln p_{ij} \quad \text{for } \mathit{all} \ j \tag{3.11}$$

where α is a constant defined as

$$\alpha = 1/ln(n)$$

which guarantees that $0 \le E_j \le 1$.

If no preferences are available *a priori*, then the best weights, instead of the equal weights, are given by

$$w_j = \frac{d_j}{\sum_{i=1}^{k} d_i} \quad \text{for } \mathit{all} \ j \tag{3.12}$$

where $d_j = 1 - E_j$ is the degree of diversity of the information involved in the outcomes of attribute j.

If the designer has *a priori*, subjective weight $\hat{w}_j$, then it can be combined with w_j, resulting in the following new weight $\tilde{w}_j$,

$$\tilde{w}_j = \frac{\hat{w}_j w_j}{\sum_{i=1}^{k} \hat{w}_i w_i} \tag{3.13}$$

The entropy method provides an alternative method of assigning weights. The subjective weight $\hat{w}_j$ may be obtained using the direct assignment method or other methods. It may be noted that the value of w_j depends on the set of alternative designs. In fact, w_j reflects the degree to which the *jth* attribute contributes in discriminating over the set of alternatives concerned. For example, an attribute does not contribute much when the attribute has similar outcomes for all alternatives. In other words, w_j can only be used to evaluate the alternatives which are used to calculate w_j.

To demonstrate the entropy method, take for example the following decision matrix about the evaluation of four designs on the basis of six attributes.

$$
\begin{array}{cc}
 & \begin{array}{cccccc} f_1 & f_2 & f_3 & f_4 & f_5 & f_6 \end{array} \\
M = \begin{array}{c} a_1 \\ a_2 \\ a_3 \\ a_4 \end{array} & \begin{bmatrix} 2.0 & 1.5 & 2.0 & 5.5 & 5.0 & 9.0 \\ 2.5 & 2.7 & 1.8 & 6.5 & 5.0 & 5.0 \\ 1.8 & 2.0 & 2.1 & 4.5 & 7.0 & 7.0 \\ 2.2 & 1.8 & 2.0 & 5.0 & 5.0 & 5.0 \end{bmatrix}
\end{array}
$$

Using formula (3.10), the attributes are normalised as follows

$$
\begin{array}{cc}
 & \begin{array}{cccccc} f_1 & f_2 & f_3 & f_4 & f_5 & f_6 \end{array} \\
[p_{ij}] = \begin{array}{c} a_1 \\ a_2 \\ a_3 \\ a_4 \end{array} & \begin{bmatrix} 0.2353 & 0.1875 & 0.2532 & 0.2558 & 0.2500 & 0.3462 \\ 0.2941 & 0.3375 & 0.2278 & 0.3023 & 0.1500 & 0.1923 \\ 0.2118 & 0.2500 & 0.2658 & 0.2093 & 0.3500 & 0.2692 \\ 0.2588 & 0.2250 & 0.2532 & 0.2326 & 0.2500 & 0.1923 \end{bmatrix}
\end{array}
$$

The entropy E_j and the weight w_j of attribute j are calculated using equations (3.11) and (3.12) as follows

$$
\begin{aligned}
E &= [E_1 \quad E_2 \quad E_3 \quad E_4 \quad E_5 \quad E_6] \\
&= [0.9946 \quad 0.9829 \quad 0.9989 \quad 0.9931 \quad 0.9703 \quad 0.9770]
\end{aligned}
$$

$$
\begin{aligned}
W &= [w_1 \quad w_2 \quad w_3 \quad w_4 \quad w_5 \quad w_6] \\
&= [0.0649 \quad 0.2055 \quad 0.0133 \quad 0.0829 \quad 0.3570 \quad 0.2764]
\end{aligned}
$$

From the above results, one can find that the weight of an attribute is small when all the alternatives have similar outcomes on the attribute. If the decision maker has *a priori* weights as given by equation (3.2), then the new weights can be obtained using (3.13) as follows

$$\begin{aligned} W &= [w_1 \quad w_2 \quad w_3 \quad w_4 \quad w_5 \quad w_6] \\ &= [0.1209 \quad 0.0851 \quad 0.0055 \quad 0.0515 \quad 0.2216 \quad 0.5154] \end{aligned}$$

3.2.4 Minimal Information Method

3.2.4.1 *General Pairwise Comparisons* and *Minimal Information*

The eigenvector method requires exact comparisons in that the DM is required to exactly evaluate how many times one attribute is more or less important than another. In engineering design, the DM may only be capable of providing a combination of exact and vague pairwise comparisons. For instance, he may assert that one attribute is *at least* twice more important than another. The minimal information trade-off assessment (MITA) method presented in [Gabbert and Brown 1989][White, Sage et al. 1984] can accommodate both exact and vague pairwise comparisons and therefore it provides a more flexible way of acquiring and representing preference information. To assign weights, the MITA method can use as much preference information as the DM can provide. Unfortunately, it does not either define the minimal information requirement formally or provide a systematic way to guide the DM in preparing his preference information.

This section introduces a new technique for weight assignment, which uses exact and/or vague pairwise comparisons of attributes for preference acquisition. It adopts an iterative procedure to assign weights, which is composed of two main steps. First, it generates an initial weight assignment based on minimum number of complete pairwise comparisons which may represent the DM's initial overall preference structure. A linear programming model is designed to facilitate the assignment. Then the initially assigned weights can be revised if the DM is not satisfied with them and if he can provide more useful information. In the procedure, the consistency and determinancy of the given comparisons are iteratively checked and numerically measured so that the DM can clearly judge the quality of the given preference information and the assigned weights. He can also judge the potential benefits of providing further pairwise comparisons of attributes. To implement the iterative procedure, a goal programming model is designed.

In the eigenvector method, the comparison matrix is assumed to be reciprocal, i.e., $m_{lh}=1/m_{hl}$, and n(n-1)/2 pairwise comparisons have to be provided. It is argued that much of the data may be redundant and furthermore this can lead to inconsistency. The Geometric Least Square (GLS) method [Islei and Lockett 1988] requires much less data. For design selection, a weight assignment technique may be more favorable if it only requires the DM to provide as much significant preference information as necessary through a systematic procedure.

The pairwise comparisons required by the eigenvector and the GLS methods are all exact ones. However, a combination of exact and vague but practical pairwise comparisons may be the best that can be provided. Suppose "*COST*" and "*FLEXIBILITY*" are two attributes, for example, the DM may describe the relative importance of the attributes using the following statements which are either exact or vague,

- *COST is the most important attribute (vague);*
- *it is better for the OPERATING COST to remain low at the expense of a higher INITIAL COST (vague);*
- *COST is twice as important as FLEXIBILITY (exact);*
- *high FLEXIBILITY and low COST are equally important (exact); and*
- *COST is at least as important as FLEXIBILITY (vague).*

In the MITA method, set inclusion is used to define the information represented by these statements. Actually, these preference statements provide information which may be transformed into linear equality or inequality constraints on the weights. For example, the "more important than" relation for attributes is a "greater than" relation on the weights; "equivalence" relation for attributes is an "equality" relation on the weights, and the "at least as important as" relation for attributes is a "greater than or equal to" relation on the weights. The "c times more important than" relation for attributes is a "equality" relation on the weights where the weight of the less important attribute is multiplied by c. Such transformation provides a set of mappings from preference statements into weight constraints.

The MITA method searches for a specific value for the weight vector W as the solution to the following mathematical programming problem

$$\begin{aligned} \min \quad & H(W) = \sum_{i=1}^{n} w_i \log \frac{w_i}{\hat{w}_i} \\ s.t. \quad & W \in \Lambda \quad \sum_{i=1}^{n} w_i = 1 \end{aligned} \tag{3.14}$$

where Λ is the feasible space defined by the weight constraints, in which W must lie. $\hat{W}=[\hat{w}_1 \cdots \hat{w}_n]^T$ is a prior preference structure or a uniform preference structure if no prior structure exists. The above model as defined by (3.14) is a nonlinear programming problem with a nonlinear objective function and generally linear constraints. The objective function $H(W)$ in (3.14) is the relative entropy, measuring the amount of information given by the preference structure W relative to $\hat{W}$. It is argued that relative entropy satisfies some inferential properties [Gabbert and Brown 1989]. For instance, if no preference information is provided by the DM (i.e. Λ is empty), equal weight on each attribute may be used by the experts, that is, W is given by $w_i=1/n$, $i=1, \cdots, n$. If a constraint on W is given, for example, $w_j \geq 2w_i$ for $i \neq j$, then $w_j = 2/(n+1)$ and $w_i=1/(n+1)$ for $i=1, \cdots, n$ and $i \neq j$ may be used.

In the MITA method, however, no systematic procedure is designed to acquire preference statements from the DM. In other words, the DM may be allowed to provide preference statements for attributes in a random manner. However, it may be better if the DM can follow a systematic yet flexible way to present preference statements, in which the consistency of the statements can be checked. Moreover, when the statements are transformed into constraints on weights which constitute Λ, the determinancy of Λ should be measured as well.

The new method, referred to as the MInimal PAirwise Comparison (MIPAC) method [Sen and Yang 1994a], uses a systematic procedure to acquire and represent preference information. In this method, relative weights of attributes can be initially assigned on the basis of a minimal number of exact and/or vague pairwise comparisons. Then the minimum comparison set may be revised or extended if the DM is not satisfied with the initial weight assignment and if he can provide more useful information. The minimum set of complete comparisons is defined as follows.

Definition 1: Suppose there are n attributes. The minimum set of complete pairwise comparisons for the n attributes is composed of (n-1) pairwise comparisons, in which each of the n attributes must be compared with at least one of the other attributes and no single comparison or a subset of the comparisons is isolated from the other comparisons.

Let y_i and y_j be two attributes and R_{ij} the preference relation for y_i and y_j. Then, definition 1 may be interpreted graphically. Suppose a circle marked by y_i represents a node. If y_i and y_j are directly compared, then the node for y_i and that for y_j are linked together by a line segment marked by R_{ij}. Thus, a direct comparison for y_i and y_j may be depicted as in Figure 3.6. The set of pairwise comparisons for n attributes may then be represented using a network, composed of n nodes and many line segments linking these nodes.

The minimum set of complete pairwise comparisons for n attributes is composed of n nodes and (n-1) line segments. Each node is directly linked with at least another node. From any node to each of the other nodes there is one and only one uninterrupted chain connected by the line segments between the nodes. The two attributes at two nodes may not necessarily be compared explicitly. They can be compared with each other either directly if there is only one line segment between them, or indirectly if there is a chain of line segments between them and if all preference relations on the chain are transitive. For instance, Figure 3.7 shows a simple minimum set of complete comparisons for three attributes, y_1, y_2, y_3. Figure 3.8 shows a more complicated minimum set of complete comparisons for nine attributes, y_i $(i=1, \cdots, 9)$.

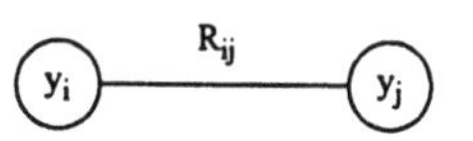

Figure 3.6 *Direct Comparison*

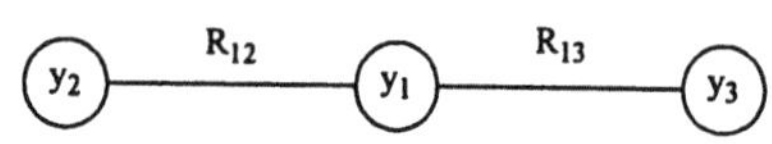

Figure 3.7 *Basic Comparison Set*

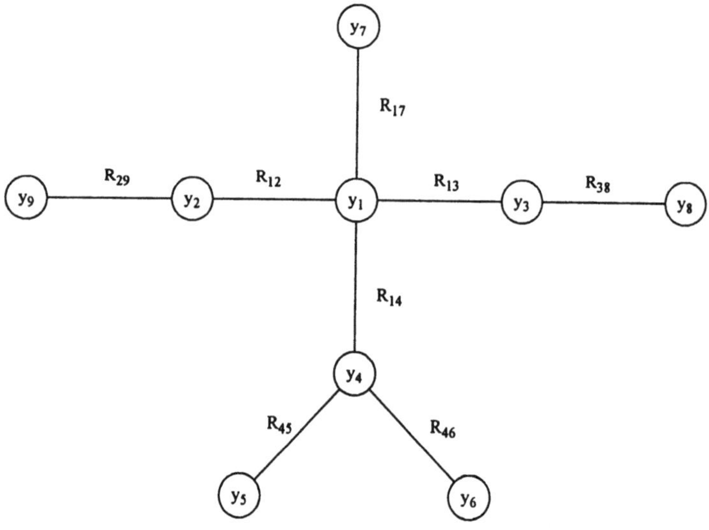

Figure 3.8 *Complex Comparison Set*

In Figure 3.7, y_1 is directly compared with y_2 and y_3 respectively. y_2 and y_3 are linked together through y_1. They may be compared indirectly or may not be compared yet, dependent upon whether or not the preference relations R_{12} and R_{13} are transitive. If the DM states that "y_2 *is exactly three times as important as* y_1" and "y_1 *is at least as important as* y_3", for example, then y_2 has actually been compared with y_3 indirectly, that is "y_2 *is at least three times as important as* y_3". If the DM states that "y_2 *is at least three times as important as* y_1" and "y_3 *is also at least three times as important as* y_1", it may not be appropriate to conclude that "y_2 *is as important as* y_3". In fact, y_2 may be more or less important than y_3. In this case, more information is necessary if the exact preference relation between y_2 and y_3 has to be determined.

It may be noted that there is more than one way to construct the minimum set of complete pairwise comparisons if there are more than two attributes. A single-chain set and a star-shaped set may be two of the simplest minimum sets. In the first set, each of the attributes is compared with at least one but at most two other attributes; in the second set, one attribute is used as the reference attribute, with which the other attributes are all compared. Other types of minimum sets may be spanned based on these two basic sets. For instance, the minimum set shown in Figure 3.7 may be regarded as a single-chain set or a star-shaped set with y_1 as the reference attribute. In Figure 3.8, if only y_i (i=1, $\cdots$, 4) is considered, the minimum set for these four attributes is star-shaped with y_1 as the reference attribute. As a whole, the minimum set shown in Figure 3.8 is a complex set with y_1, y_2, y_3 and y_4 as the reference attributes.

3.2.4.2 *Linear Programming Models* for *Weight Assignment*

Once the minimum set of complete pairwise comparisons for attributes is built, these comparisons can be transformed into linear equality or inequality constraints on the weights, which constitute the minimum constraint set $\Lambda_{\min}$. Generally, $\Lambda_{\min}$ may be written as follows

$$\Lambda_{\min} = \left\{ W \;\middle|\; \begin{array}{l} W=[w_1 \; \cdots \; w_n]^T, \; c_i w_i \Delta_r c_j w_j \text{ for } r=1,\cdots,n-1, \\ \textit{where } i,j \in \{1,\cdots,n\}, \; i \neq j; \; \Delta_r \in \{<, >, =, \leq, \geq\}; \\ \text{and } c_i \text{ and } c_j \textit{ are real numbers} \end{array} \right\} \tag{3.15}$$

According to Definition 1, $\Lambda_{\min}$ is not empty.

To avoid using a nonlinear function as a standard for weight assignment such as the relative entropy as defined in problem (3.14), a p-norm function is employed for assigning the best compromise weight. The p-norm function is defined as follows

$$\| W^* - W \|_p = \left[\sum_{i=1}^{n} (w_i^* - w_i)^p \right]^{1/p} \tag{3.16}$$

where p is positive and $W^*=[w_1^* \cdots w_n^*]^T$ is the ideal weight vector with w_i^* being the ideal weight for the attribute y_i. If the weight vector is normalized, that is $\sum_{i=1}^n w_i = 1$, then let $w_i^*=1$ $(i=1, \cdots, n)$ as the maximum possible weight for each attribute is one. It is easy to show that the p-norm also possesses discriminating characteristics of the relative entropy method.

It is always desirable that the best compromise weight is assigned to be as close as possible to the ideal weight. The mathematical programming problem for initial weight assignment may thus be formulated as follows

$$\begin{cases} \min & \| W^* - W \|_p \\ s.t. & W \in \Lambda \end{cases} \tag{3.17}$$

where

$$\Lambda = \{ W \mid W \in \Lambda_{\min};\ \sum_{i=1}^{n} w_i=1;\ w_i \geq 0;\ i=1, \cdots, n \}$$

Λ represents the feasible region for weight assignment, in which there is at least one feasible solution. The optimal solution of (3.17) may be used as the best compromise weight vector which is nearest W^* in the sense of p-norm. Let $p = \infty$. The ∞-norm may then be used to search for the best compromise weight as the problem (3.17) with $p = \infty$ can be transformed into the following minimax problem, i.e.

$$\begin{cases} \min & \| W^* - W \|_\infty \\ s.t. & W \in \Lambda \end{cases} \quad \Leftrightarrow \quad \min_{W \in \Lambda} \max_i \{(w_i^* - w_i)\} \tag{3.18}$$

and (3.18) has the following equivalent

$$\begin{aligned} & \min \ \lambda \\ & s.t. \quad w_i^* - w_i \leq \lambda \quad i=1, \cdots, n; \quad W \in \Lambda, \quad \lambda \geq 0 \end{aligned} \tag{3.19}$$

which is only a linear programming problem.

If there is exactly one feasible solution in Λ, which is the case when the relations Δ_r in $\Lambda_{\min}$ (see (3.15)) are all exact ones (i.e., Δ_r is "=" for all $r=1, \cdots, n-1$), then the best compromise weight vector is precisely determined by the DM's preference statements.

If there is more than one feasible solution in Λ, which is generally the case, the best compromise weight vector is under-determined and may be generated as the optimal solution of (3.19). However, other feasible solutions in Λ may also be selected as the best compromise weight vector by the DM if he is not satisfied with the optimal solution of (3.19) and if there exist other solutions in Λ which are significantly different from and better than the current optimum. Hence it may be useful to define a measure to check the determinancy of the DM's preference statements, so that the DM can clearly know how much room remains for weight assignment.

A determinancy index (simply, *DI*) is then defined as follows. Suppose $\overline{W}^{(j)} = [\overline{w}_1^{(j)} \cdots \overline{w}_n^{(j)}]^T$ is the optimal solution of the following problem

$$\max_{W \in \Lambda} w_j \qquad j=1, \cdots, n \tag{3.20}$$

$\overline{W}^{(j)}$ is called an extreme weight vector and $\overline{w}_j^{(j)}$ is the maximal feasible weight value for the attribute y_j. The area of the feasible weight vectors on the normalization hyperplane (Λ) may be a measure to indicate the determinancy, although any other measure can be conveniently substituted. As this area is difficult to calculate, the area of the hyperpolygon enclosed by connecting the extreme weight vectors on the normalization hyperplane may be used to approximate the whole feasible area. As the feasible area is a convex set, the constructed hyperpolygon is always part of it.

Define $E(\overline{W})$ as the mean vector of the n extreme weight vectors, that is

$$E(\overline{W}) = \frac{1}{n}\sum_{j=1}^{n} \overline{W}^{(j)} = [E(\overline{w}_1) \cdots E(\overline{w}_n)]^T \tag{3.21}$$

$$E(\overline{w}_i) = \frac{1}{n}\sum_{j=1}^{n} \overline{w}_i^{(j)} \qquad i=1, \cdots, n \tag{3.22}$$

Obviously, $E(\overline{W})$ is the geographical centre of the hyperpolygon. Then define a normalized Euclidian distance between the mean weight vector $E(\overline{W})$ and the jth extreme weight vector as follows

$$D_j = \left[\frac{\sum_{i=1}^{n} (\overline{w}_i^{(j)} - E(\overline{w}_i))^2}{n(n-1)}\right]^{1/2} \qquad j=1, \cdots, n \tag{3.23}$$

where the denominator $n(n-1)$ is a scaling factor. The *DI* may then be defined by

$$DI = 1 - \sum_{j=1}^{n} D_j \tag{3.24}$$

It is easy to prove that $DI \in [0 \;\; 1]$ if $\overline{W}^{(j)}$ is generated using (3.20). When there is only one solution in Λ, $DI = 1$ as $\overline{W}^{(1)} = \cdots = \overline{W}^{(n)} = E(\overline{W}) = \hat{W}$ where $\hat{W}$ is the unique pre-determined weight vector in Λ. $DI = 0$ when no specific preference information is provided except for the normalizing constraint on weights, i.e., $\Lambda_{min} = \varnothing$ in (3.15). In fact, it is easy to calculate from (3.19) that in this case $\overline{w}_i^{(j)} = 0$ $(i=1, \cdots, n;\ i \neq j)$ and $\overline{w}_j^{(j)} = 1$ $(j=1, \cdots, n)$ where $\overline{W}^{(j)} = [\overline{w}_1^{(j)} \cdots \overline{w}_n^{(j)}]^T$. So, $E(\overline{W}) = [1/n \cdots 1/n]^T$ and $D_j = 1/n$, $j=1, \cdots, n$.

Figures 3.9 to 3.12 demonstrate a weight assignment problem for three attributes y_1, y_2 and y_3 with four sets of preference statements. It may be noted that the same best compromise weight vector $\hat{W} = [1/3 \;\; 1/3 \;\; 1/3]^T$ can be obtained for the four sets of statements using (3.19). However, the determinacies of the four sets of statements are different.

In Figure 3.9, no preference information is provided, so that any solution on the normalization plane ($w_1 + w_2 + w_3 = 1$; w_1, w_2, $w_3 \geq 0$) might be selected to be the best weight vector. The three extreme weight vectors are $\overline{W}^{(1)} = [1 \;\; 0 \;\; 0]^T$, $\overline{W}^{(2)} = [0 \;\; 1 \;\; 0]^T$, $\overline{W}^{(3)} = [0 \;\; 0 \;\; 1]^T$, and the corresponding mean weight vector is $E(\overline{W}) = [1/3 \;\; 1/3 \;\; 1/3]^T$. The value of DI^1 is then zero. In Figure 3.12, a minimum set of two exact complete comparisons is provided. Since $\hat{W} = \overline{W}^{(1)} = \overline{W}^{(2)} = \overline{W}^{(3)} = E(\overline{W}) = [1/3 \;\; 1/3 \;\; 1/3]^T$, $DI^4 = 1$. In Figure 3.10, only one comparison is provided, which is incomplete, and $DI^2 = 0.21$. In Figure 3.11, a minimum set of two vague complete comparisons is provided and $DI^3 = 0.5$.

In Figure 3.9, the area of the feasible weight vectors, the shaded area, is the same as the polygon (triangle) enclosed by connecting the three extreme weight vectors. This is also the case in Figures 3.10 and 3.12. In Figure 3.11, the latter is enclosed by the former as the three extreme weight vectors (points) are (1/2, 0, 1/2), (1, 0, 0) and (1/2, 1/2, 0). The triangle enclosed by connecting these three points is within the shaded area.

From the illustrative examples, it is obvious that a larger value of DI indicates that the quality of preference information is better. Preferences with better quality can more precisely determine the best weight vector but they are more difficult to provide. If DI is large enough (near one), it makes no sense to acquire more preference statements from the DM because in this case other feasible weight vectors in Λ are not significantly different from the optimum of (3.19), and there is some danger of providing information that is inconsistent with that already given. If the value of DI is not good enough, more information may be required so as either to revise the existing comparisons in the minimum set or to compare more attributes directly. In this latter case too, the added direct comparisons may be inconsistent with the ones in the minimum set. It is therefore necessary to check the consistency of the added comparisons.

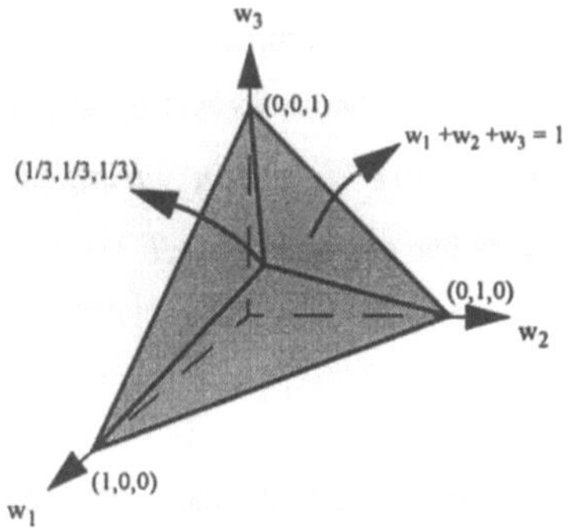

Figure 3.9 *No Preference Infomation*

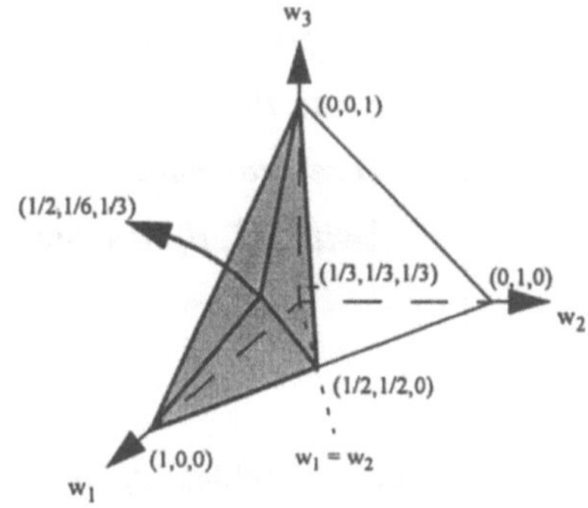

Figure 3.10 *One Comparison* $w_1 / w_2 >= 1$

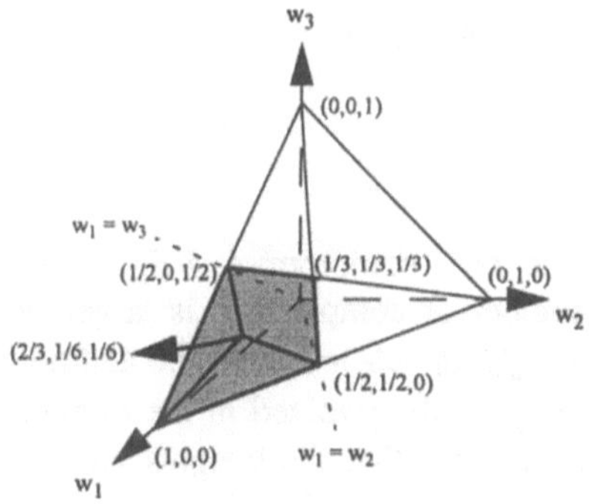

Figure 3.11 *Comparisons* $w_1 / w_2 >= 1$ *and* $w_1 / w_3 >= 1$

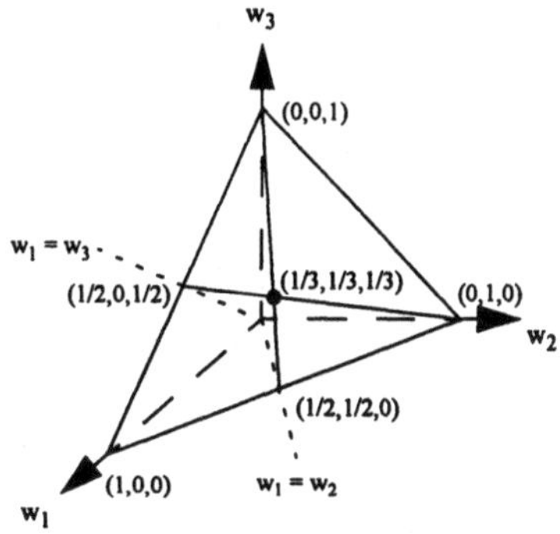

Figure 3.12 *Comparisons* $w_1 / w_2 = 1$ *and* $w_1 / w_3 = 1$

Suppose Λ_a is an additional sub-set, which is composed of the added direct comparisons except for those involved in the minimum set and is defined as follows

$$\Lambda_a = \left\{ W_a \left| \begin{array}{l} W_a=[w_1 \ \cdots \ w_n \ d_1^+ \ d_1^- \ \cdots \ d_T^+ \ d_T^-]^T \\ a_i w_i \Delta_t a_j w_j + d_t^+ - d_t^- \text{ for } t=1, \cdots, T \\ where \ i, j \in \{1, \cdots, n\}, \ i \neq j; \ \Delta_t \in \{<, >, =, \leq, \geq\} \\ T \ is \ the \ number \ of \ the \ additional \ comparisons; \ a_i \text{ and } a_j \\ are \ real \ numbers; \text{ and } d_t^+ \text{ and } d_t^- \ are \ deviation \ variables \\ with \ d_t^+, d_t^- \geq 0 \text{ and } d_t^+ \times d_t^- = 0 \text{ for } all \ t=1, \cdots, T \end{array} \right. \right\} \tag{3.25}$$

where deviation variables d_t^+ and d_t^- measure the consistency of the added comparisons with those in $\Lambda_{\min}$. The best compromise weight vector is then assigned using the following linear goal programming formulation, where $P_1 \gg P_2$

$$\begin{aligned} &\min \ \{P_1 \sum_{t=1}^{T} (d_t^+ + d_t^-) + P_2 \| W^* - W \|_\infty\} \\ &s.t. \quad W_a \in \Lambda_a, \ W \in \Lambda \end{aligned} \tag{3.26}$$

As a whole, the MIPAC method assigns the best compromise weight vector using the following two main steps if the number of comparisons is larger than $(n-1)$. At first, the consistency is checked. If $\sum_{t=1}^{T}(d_t^+ + d_t^-)=0$, then the additional pairwise comparisons are consistent with those already involved in the minimum set. Otherwise, inconsistency occurs, which indicates that the weights are over-determined. The inconsistent comparisons with d_t^+ or d_t^- being greater than zero can then be identified. The DM may either revise these comparisons or the relevant comparisons in the minimum set. Then, the best compromise weight vector is assigned to be the solution which is nearest to the ideal weight vector in the sense of ∞-norm, or in a minimax sense.

The MIPAC method thus provides a flexible and systematic procedure to acquire preference information. It initially requires the DM to provide a minimum number of complete pairwise comparisons for attributes so as to generate the first weight assignment using (3.19). If the DM is not satisfied with the initially assigned weights and if he can provide more useful preference information, the method will ask the DM either to revise the existing comparisons in the minimum set or to take into account more direct comparisons so that better compromise weights can be assigned. The consistency and determinancy of the comparisons can be checked and numerically measured so that the DM clearly knows the quality of the preference information he has provided and hence the quality of the compromise weights he has obtained.

3.2.4.3 *An Example*

A mathematical model for preliminary design of a semi-submersible has been built as a multiple objective decision making (MODM) problem, which can be generalized as the following vector nonlinear programming problem

$$\begin{aligned} &\max \ \{y_1(a) \ y_2(a) \ y_3(a) \ y_4(a) \ y_5(a)\} \\ &\min \ \{y_6(a)\} \\ &s.t. \ \ g_i(a) \le 0 \qquad i=1, \cdots, 11 \\ &\qquad h_i(a) \le 0 \qquad j=1, \cdots, 9 \qquad a = [x_1 \ x_2 \ \cdots \ x_9]^T \end{aligned} \tag{3.27}$$

where $y_i(a)$ $(i=1, \cdots, 6)$ are nonlinear objective functions, $g_i(a)$ $(i=1, \cdots, 11)$ are nonlinear constraint functions and $h_i(a)$ $(i=1, \cdots, 9)$ are linear constraint functions. The objective functions and the design variables are described in Table 3.2. The purpose of design is to generate a best compromise design which can attain the best possible values for these six objectives.

Table 3.2 *Description of Semi-Submersible Model*

design variables		objective functions		
symbol	description	symbol	units	description
x_1	corner column diameter	y_1	seconds	natural heave period
x_2	middle column diameter	y_2	tonnes	transit payload
x_3	length of the column	y_3	tonnes	operating payload
x_4	breadth of the pontoon	y_4	meters	permissible KG in transit
x_5	depth of the pontoon	y_5	meters	permissible KG in operation
x_6	length of the pontoon	y_6	pounds	cost of steel construction
x_7	height of the deck			
x_8	distance between pontoon centrelines			
x_9	diameter of transverse bracing			

Since there is no single design which could optimize (maximize or minimize) the six objectives simultaneously, compromise analysis among the objectives is necessary, based on the DM's preference information about the relative importance of the objectives. If preference information is acquired and represented by a utility function, the best compromise design may be obtained by optimizing the utility function. In this section, an alternative design synthesis strategy is presented. Firstly, an interactive MODM method is used to generate a set of efficient (non-dominated) designs. Then, the MIPAC method is used to assign weights for the objectives (attributes). The best compromise design can then be selected from the generated efficient designs using a MADM method.

The interactive step trade-off method (ISTM) is used to generate the efficient designs [Yang et al. 1988, 1990], as described in the next chapter. The interactive efficient design generation process is illustrated in [Yang 1992c]. Table 3.3 lists the values of the six objectives at the 13 generated efficient designs. The "-" symbol in the last column in Table 3.3 means that y_6 is for minimization. The first six designs ($a_1 \cdots a_6$) are the extreme designs (efficient ones) generated by optimizing each of the six objective functions separately. The values of the design variables for the extreme designs are shown in Table 3.4. The design a_{10} is referred to as the feasible ideal design which is closest to the imaginary ideal design taking the best feasible value of each objective. This feasible ideal design is generated assuming that all the objectives are of equal importance. The other six efficient designs are generated near the feasible ideal design using an interactive decision making procedure. The remaining problem is then to rank these 13 designs by taking the DM's preferences into account.

Table 3.3 *The Decision Matrix* for *The Semi-Submersible*

efficient	objective values					
designs	y_1	y_2	y_3	y_4	y_5	y_6
a_1	36.06	9594.95	18014.24	18.33	18.31	-7594448.00
a_2	32.64	13040.96	26194.62	21.33	21.34	-12404110.00
a_3	31.26	12284.91	30248.64	25.92	25.93	-15256452.00
a_4	28.21	6952.27	20811.53	32.29	32.32	-13551909.00
a_5	28.23	6748.00	20270.23	32.29	32.32	-13205347.00
a_6	21.87	2288.60	7608.83	20.36	20.35	-4805733.50
a_7	31.48	7892.82	16317.26	23.11	23.10	-7848295.50
a_8	31.48	7663.19	15785.09	22.11	22.10	-7430433.50
a_9	29.50	6663.20	14785.06	23.38	23.38	-7430433.00
a_{10}	31.48	8448.19	20186.60	26.11	26.09	-9450440.00
a_{11}	31.48	8312.49	19585.60	25.61	25.59	-9078700.00
a_{12}	31.61	8311.80	19584.40	25.21	25.19	-8946110.00
a_{13}	31.01	7985.11	17584.40	25.21	25.21	-8659400.00

Table 3.3 actually provides numerical values for multiple attribute evaluations of the efficient designs generated. If there was a design attaining the best values for all the six attributes, it would of course be selected as the best design. Unfortunately, such a design does not exist for the problem as some of the objectives are in conflict. Thus the ranking of the efficient designs depends not

Table 3.4 *Variable Values of the Extreme Designs*

design variables	extreme designs					
	a_1	a_2	a_3	a_4	a_5	a_6
x_1	12.59	15.00	15.00	14.99	15.00	9.86
x_2	7.00	10.00	12.00	7.50	7.00	8.22
x_3	26.80	26.80	30.00	40.00	40.00	26.80
x_4	25.00	25.00	25.00	19.99	20.00	13.15
x_5	8.93	8.93	10.00	10.00	10.00	8.93
x_6	117.53	150.00	150.00	132.18	128.98	90.83
x_7	4.00	7.07	10.00	10.00	10.00	4.00
x_8	75.53	90.00	90.00	89.96	90.00	59.17
x_9	2.69	1.00	1.00	1.00	1.00	1.00

only on the multiple attribute evaluations as given in Table 3.3 but also on the preference information of the DM about the relative importance of the six attributes, which may be represented as weights.

The MIPAC method is used to assign the weights for the objectives. It is assumed that the DM initially provides the following pairwise comparisons for the objectives.

1> *"COST OF CONSTRUCTION (y_6)" is at least twice as important as "NATURAL HEAVE PERIOD (y_1)" (R_{16}).*

2> *"COST OF CONSTRUCTION (y_6)" is at least three times as important as "OPERATING PAYLOAD (y_3)" (R_{36}).*

3> *"PERMISSIBLE KG IN OPERATION (y_5)" is at least twice as important as "OPERATING PAYLOAD (y_3)" (R_{35}).*

4> *"OPERATING PAYLOAD (y_3)"is at least twice as important as "TRANSIT PAYLOAD (y_2)" (R_{23}).*

5> *"PERMISSIBLE KG IN TRANSIT (y_4)" is as important as "PERMISSIBLE KG IN OPERATION (y_5)" (R_{45}).*

The comparisons R_{16}, R_{36}, R_{35} and R_{23} are vague ones and the last comparison R_{45} is an exact one. The above set of comparisons can be depicted as shown in Figure 3.13. Obviously, these five comparisons constitute a minimum set of complete pairwise comparisons for the six objectives.

These preference statements are then transformed into the constraints on the weights. Suppose w_i is the relative weight for y_i, $i=1, \cdots ,6$. Then the initial minimum set $\Lambda_{\min}^{(0)}$ can be constructed as follows

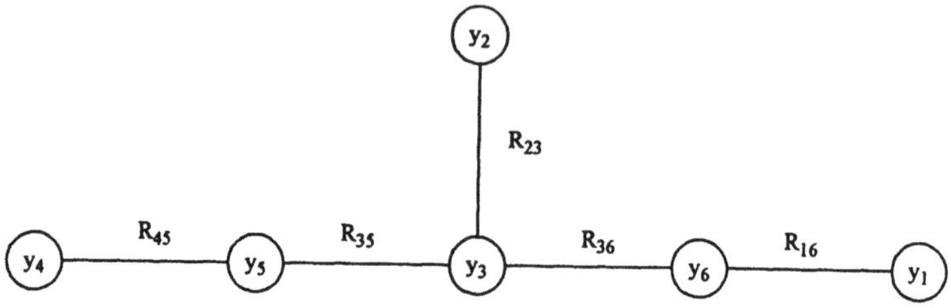

Figure 3.13 *Minimum Comparison Set for Semi-Submersible*

$$\Lambda_{\min}^{(0)} = \left\{ W \;\middle|\; \begin{array}{l} w_6 - 2w_1 \geq 0, \;\; w_6 - 3w_3 \geq 0, \;\; w_5 - 2w_3 \geq 0 \\ 2w_2 - w_3 \leq 0, \;\; w_4 - w_5 = 0, \;\; W = [w_1 \;\; \cdots \;\; w_6]^T \end{array} \right\} \tag{3.28}$$

The initial linear programming problem for assigning the weights can be constructed as follows

$$\begin{array}{ll} \min & \| W^* - W \|_\infty \\ s.t. & W \in \Lambda^{(0)} \end{array} \tag{3.29}$$

where

$$W^* = [1 \;\; \cdots \;\; 1]^T \tag{3.30}$$

$$\Lambda^{(0)} = \{ W \mid W \in \Lambda_{\min}^{(0)}, \; \sum_{i=1}^{6} w_i = 1, \; w_i \geq 0, \; i=1, \cdots, 6\} \tag{3.31}$$

or equivalently

$$\begin{array}{lll} \min & \lambda & \\ s.t. & 1 - w_1 \leq \lambda & w_6 - 2w_1 \geq 0 \\ & 1 - w_2 \leq \lambda & w_6 - 3w_3 \geq 0 \\ & 1 - w_3 \leq \lambda & w_5 - 2w_3 \geq 0 \\ & 1 - w_4 \leq \lambda & 2w_2 - w_3 \leq 0 \\ & 1 - w_5 \leq \lambda & w_4 - w_5 = 0 \\ & 1 - w_6 \leq \lambda & \sum_{i=1}^{6} w_i = 1 \\ & \lambda \geq 0, \; w_i \geq 0, \; i=1, \cdots, 6 & \end{array} \tag{3.32}$$

The optimal solution of (3.32) is $\hat{W}^{(0)} = [0.0556\ \ 0.0556\ \ 0.1111\ \ 0.2222\ \ 0.2222\ \ 0.3333]^T$ and the value of the determinancy index is $DI^{(0)} = 0.5385$. $DI^{(0)}$ is rather small, which means that much room remains for improvement of the weight assignment. It could be considered, for example, that the DM is not satisfied with the initial weight assignment in that the first objective "*NATURAL HEAVE PERIOD* (y_1)" is a very important performance index but it has been assigned the lowest weight. The DM therefore takes into consideration the following two additional comparisons.

6> *"NATURAL HEAVE PERIOD* (y_1)" *is 1.5 times as important as "PERMISSIBLE KG IN OPERATION* (y_5)" (R_{15}).

7> *"COST OF CONSTRUCTION* (y_6)" *is at most 2.5 times as important as "NATURAL HEAVE PERIOD* (y_1)" (R_{16}).

The two added comparisons can be transformed to the additional constraint subset $\Lambda_a^{(1)}$ on weights,

$$\Lambda_a^{(1)} = \left\{ W_a \;\middle|\; \begin{array}{l} w_1 - 1.5w_5 + d_1^+ - d_1^- = 0,\ w_6 - 2.5w_1 - d_2^- \le 0 \\ d_1^+ \times d_1^- = 0;\ d_1^+, d_1^-, d_2^- \ge 0 \\ W_a = [w_1 \cdots w_6\ d_1^+\ d_1^-\ d_2^-]^T \end{array} \right\} \tag{3.33}$$

Furthermore, the DM agrees that the "at least" in the comparisons R_{35} and R_{23} (statements 3> and 4>) can now be removed so that R_{35} and R_{23} become exact preference relations instead of the original vague ones. The initial minimum set $\Lambda_{\min}^{(0)}$ is thus revised to be $\Lambda_{\min}^{(1)}$,

$$\Lambda_{\min}^{(1)} = \left\{ W \;\middle|\; \begin{array}{l} w_6 - 2w_1 \ge 0,\ \ w_6 - 3w_3 \ge 0,\ \ w_5 - 2w_3 = 0 \\ 2w_2 - w_3 = 0,\ \ w_4 - w_5 = 0,\ \ W = [w_1\ \cdots\ w_6]^T \end{array} \right\} \tag{3.34}$$

The linear goal programming for improving the initial weight assignment is then formulated by

$$\begin{array}{ll} \min & \{P_1[(d_1^+ + d_1^-) + d_2^-] + P_2 \| W^* - W \|_\infty\} \\ s.t. & W_a \in \Lambda_a^{(1)},\ W \in \Lambda^{(1)} \end{array} \tag{3.35}$$

where

$$\Lambda^{(1)} = \{ W \mid W \in \Lambda_{\min}^{(1)};\ \sum_{i=1}^{6} w_i = 1,\ w_i \ge 0,\ i=1,\cdots,6\} \tag{3.36}$$

Solving (3.35), the new optimum $\hat{W}^{(1)} = [0.2069 \;\; 0.0345 \;\; 0.069 \;\; 0.1379 \;\; 0.1379 \;\; 0.4138]^T$ can be obtained with $d_1^+ = d_1^- = d_2^- = 0$ and $DI^{(1)} = 0.9784$. Thus, the added comparisons R_{15} and R_{16} are consistent with those listed in the revised minimum set $\Lambda_{\min}^{(1)}$. $DI^{(1)}$ is now large enough and $\hat{W}^{(1)}$ may be used as the best compromise weight vector, that is,

$$W = \hat{W}^{(1)} = [0.2069 \;\; 0.0345 \;\; 0.069 \;\; 0.1379 \;\; 0.1379 \;\; 0.4138]^T \quad (3.37)$$

If the DM is not satisfied with $\hat{W}^{(1)}$ either, he may further revise the minimum set and/or provide more direct comparisons. For instance, the DM may add that

8> *"COST OF CONSTRUCTION* (y_6)*" is at most 7 times as important as "OPERATING PAYLOAD* (y_3)*"* (R_{36}).

Therefore, $\Lambda_a^{(1)}$ defined in (3.33) is changed into

$$\Lambda_a^{(2)} = \left\{ W_a \;\middle|\; \begin{array}{l} w_1 - 1.5w_5 + d_1^+ - d_1^- = 0 \\ w_6 - 2.5w_1 - d_2^- \le 0 \\ w_6 - 7w_3 - d_3^- \le 0 \\ d_1^+ \times d_1^- = 0; \;\; d_1^+, d_1^-, d_2^-, d_3^- \ge 0 \\ W_a = [w_1 \cdots w_6 \; d_1^+ \; d_1^- \; d_2^- \; d_3^-]^T \end{array} \right\}$$

Solving the following new linear goal programming problem

$$\begin{array}{ll} \min & \{P_1[(d_1^+ + d_1^-) + d_2^- + d_3^-] + P_2 \| W^* - W \|_\infty\} \\ s.t. & W_a \in \Lambda_2^{(2)}, \;\; W \in \Lambda^{(1)} \end{array} \quad (3.38)$$

we obtain the optimal solution $\hat{W}^{(2)}$. The optimal value of $\hat{W}^{(2)}$ is equal to $\hat{W}^{(1)}$ with $d_1^+ = d_1^- = d_2^- = d_3^- = 0$ and $DI^{(2)} = 0.985 > DI^{(1)}$. So the added direct comparison for y_3 and y_6 has improved the determinancy or quality of the preference information.

On the other hand, if the statement 8> is replaced by

8-1> *"COST OF CONSTRUCTION* (y_6)*" is at most 5 times as important as "OPERATING PAYLOAD* (y_3)*"* (R_{36}).

It can be shown that (R_{36}) is inconsistent with the statements 1> to 7>, that is $d_3^- > 0$. As a matter of fact, the statements 1> to 7> imply that $7.5 \ge w_6/w_3 \ge 6$.

3.3 Typical MADM Methods and Applications

In Section 3.1, we discussed how to classify MADM methods. As a result of the discussion, four MADM methods have been selected for the development of a decision support system. These methods are the AHP method, the UTA method, the TOPSIS method and the CODASID method. In this section, the computation procedures of these methods are to be described and the application examples of the methods will also be demonstrated. More detailed description of these methods can be found in the references.

3.3.1 AHP Method and Application

Saaty proposed the Analytic Hierarchy Process (AHP) to deal with MADM problems possibly with hierarchical structures of attributes as shown in Figure 3.2 [Saaty 1988]. In a hierarchical MADM problem, at the top level there is only a single element, that is the overall goal of the problem, and each element at a given level is associated with some or all of the elements at the level immediately below. Elements at a single level are compared in terms of relative attractiveness or importance with respect to an element in the immediate higher level. Such pairwise comparisons are then treated using the eigenvector method. The process is repeated from the top to the bottom of the hierarchy. The purpose of the process is to rank all the elements at a single level with respect to the top element.

The AHP method provides a simple way to formulate a MADM problem and to elicit preference information as it only requires comparisons between attributes or alternatives. The computational steps of AHP may be summarized as follows.

Step 1: Establish the hierarchy of a MADM problem as shown in Figures 3.1 and 3.2.

Step 2: Formulate a pairwise comparison matrix for elements at a single level of the hierarchy with respect to each of the elements at a level immediately above.

Step 3: Define the weight of the unique element at top level as w^1. Generate the priority vector for elements at the second level using the eigenvector method, denoted by b_{21}. In other words, b_{21} is the normalized eigenvector of the corresponding pairwise comparison matrix with respect to the largest eigenvalue. Then the relative weight vector of the elements at the second level is

$$w^2 = b_{21}w^1 = B_2 w^1 \tag{3.39}$$

where

$$B_2 = b_{21} \tag{3.40}$$

Step 4: Suppose b_{qh} is the priority vector of the elements at the *qth* level with respect to the *hth* element at the *(q-1)th* level (q>1). The priority matrix of the elements at the *qth* level can be defined as follows

$$B_q = [b_{q1} \cdots b_{qh} \ldots] \tag{3.41}$$

Then the relative weight vector of the elements at the *qth* level can be calculated as follows

$$w^q = B_q w^{q-1} \quad (q>1) \tag{3.42}$$

Step 5: Rank the elements at the *qth* level based on the relative weight vector of this level, w^q. An element with a large value of the relative weight in w^q is more favorable.

AHP is a well-known MADM methods. However, it has several weak points as discussed in reference [Stewart 1992]. For example, it implicitly assumes that elements at any single level except for the bottom level are preferentially independent. If attributes in a MADM problem are not allowed to be evaluated independently, it may be inadequate to use AHP to deal with the problem. Unfortunately, the evaluation of an attribute in a MADM problem may most probably depend upon the achievement levels of other attributes. Another weak point of AHP is that it requires that each alternative be compared with all others while many of such comparisons are redundant. This often causes an inconsistency problem. Such inconsistency may become worse as the dimension of a comparison matrix increases. The method also suffers from the rank reversal of alternatives depending on the number of alternatives being assessed and this can be a disturbing factor in a normative decision making tool. Notwithstanding these complaints, AHP does clarify many decisin situations and can lead to acceptable answers if the technique is used with caution.

Figure 3.14 shows a vessel choice problem involving the selection of one type of ship from three candidate vessel types for coastal transport in a developing country [Sen 1992]. These three vessel types are general cargo type, ro-ro type and full-container type, denoted by a_1, a_2 and a_3. The AHP method is applied to the hypothetical choice problem. It should be stated at the outset that the relative weights assigned to various factors at the different levels are based on judgements and could change depending on who is doing the assessment. Since the method allows a range of decision makers and scenarios to be conveniently included, it can cope with composite as well as personal preferences. The scheme also permits the incorporation of the relative importance of different scenarios and this facility makes it possible to incorporate the perceived likelihoods associated with the range of scenarios.

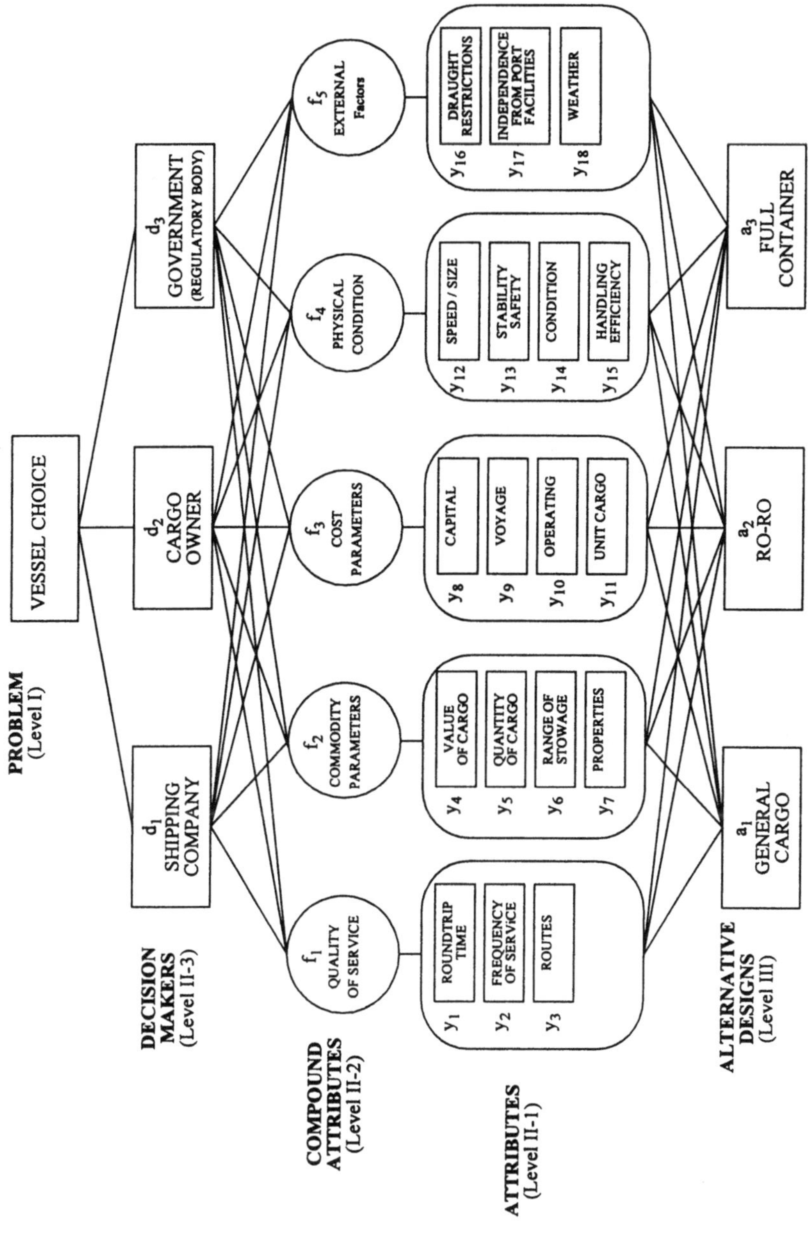

Figure 3.14 *Example of Ship Choice Problem Using the Analytic Hierarchy Process*

Three decision makers (DMs) involved in the selection of the vessel types are the shipping company, the cargo owner and the government, represented by d_1, d_2 and d_3, respectively. To evaluate these vessel types, the DMs take into account five factors (f_i, i=1, $\cdots$,5), which are measured in terms of eighteen attributes (y_j, j=1, $\cdots$,18). f_i and y_j (i=1, $\cdots$,5; j=1, $\cdots$,18) are defined in Figure 3.14. For instance, the factor "Quality of service (f_1)" is measured by means of three attributes, "Total roundtrip time (y_1)", "Frequency of service (days between calls) (y_2)" and "Route options (alternative routes) (y_3)".

Table 3.5 shows the recommended scale used for making comparisons [Saaty 1988]. When the elements being compared are closer together in preference than indicated by the scale, one can use the values 1.1, 1.2, $\cdots$, 1.9 or up to any decimal place thought to be appropriate.

Table 3.5 *Intensity of Importance*

scale	definition
1	equal importance
3	moderate importance of one over another
5	essential or strong importance
7	very strong importance
9	extreme importance
2,4,6,8	intermediate values between the two adjacent judgements

At level II-3 of Figure 3.14, if the decision makers are all equally important to the scheme then the matrix can be written as in Table 3.6-1 (Only the upper triangular part of the matrix is shown because of the reciprocal nature of the remaining elements).

Table 3.6-1 *Comparison Matrix* 1

vessel choice	d_1	d_2	d_3
d_1	1	1	1
d_2		1	1
d_3			1

$$W1 = \begin{bmatrix} 0.333 \\ 0.333 \\ 0.333 \end{bmatrix} \qquad \lambda_{max} = 3.000 \qquad CI = 0.000$$

As expected all elements of $W1$ are equal. Since all the comparisons in Table 3.6-1 are the same, they are completely consistent. We thus have CI=0.

Level II-2 represents the relative preferences of the decision makers for the five factors. For the Shipping Company (d_1), these factors are pairwisely compared as in Table 3.6-2.

Table 3.6-2 *Comparison Matrix* 2

d_1	f_1	f_2	f_3	f_4	f_5
f_1	1	1/2	1	1/3	2
f_2		1	1/2	1/3	2
f_3			1	1	3
f_4				1	2
f_5					1

$$W2 = \begin{bmatrix} 0.1497 \\ 0.1698 \\ 0.2572 \\ 0.3279 \\ 0.0954 \end{bmatrix} \qquad \lambda_{max} = 5.2898 \qquad CI = 0.0725$$

For the Shipping Company therefore, the order of importance of the five factors is f_4, f_3, f_2, f_1 and f_5. Similar results can be obtained for d_2 and d_3

Table 3.6-3 *Comparison Matrix* 3

d_2	f_1	f_2	f_3	f_4	f_5
f_1	1	1/3	2	3	2
f_2		1	3	3	2
f_3			1	4	2
f_4				1	1
f_5					1

Table 3.6-4 *Comparison Matrix* 4

d_3	f_1	f_2	f_3	f_4	f_5
f_1	1	3	4	4	1
f_2		1	2	2	1/2
f_3			1	1	1/3
f_4				1	1/3
f_5					1

$$W3 = \begin{bmatrix} 0.2253 \\ 0.3935 \\ 0.1852 \\ 0.0840 \\ 0.1120 \end{bmatrix} \qquad \lambda_{max} = 5.3396 \qquad CI = 0.0849$$

$$W4 = \begin{bmatrix} 0.3664 \\ 0.1556 \\ 0.0889 \\ 0.0889 \\ 0.3002 \end{bmatrix} \qquad \lambda_{max} = 5.0333 \qquad CI = 0.0083$$

At level II-1, for factors f_1 and f_2, there are three and four lower-level attributes. These attributes are pairwisely compared as in Table 3.6-5 and 3.6-6.

Table 3.6-5 *Comparison Matrix* 5

f_1	y_1	y_2	y_3
y_1	1	1	2
y_2		1	2
y_3			1

Table 3.6-6 *Comparison Matrix* 6

f_2	y_4	y_5	y_6	y_7
y_4	1	1/3	1	1/3
y_5		1	4	1
y_6			1	1/3
y_7				1

$$C1 = \begin{bmatrix} 0.4000 \\ 0.4000 \\ 0.2000 \end{bmatrix} \qquad \lambda_{max} = 3.0000 \qquad CI = 0.0000$$

$$C2 = \begin{bmatrix} 0.1225 \\ 0.3959 \\ 0.1142 \\ 0.3674 \end{bmatrix} \qquad \lambda_{max} = 4.0104 \qquad CI = 0.0035$$

Thus, for factor f_1, the order of importance of relevant attributes y_1 and y_2 are equal while y_3 is the least important. For factor f_2, relevant attribute y_5 is the most important and y_3 the least important. Similar results can be obtained for the other factors, as shown in Tables 3.6-7, 3.6-8 and 3.6-9.

Table 3.6-7*Comparison Matrix* 7

f_3	y_8	y_9	y_{10}	y_{11}
y_8	1	1/3	1	1/2
y_9		1	3	1/2
y_{10}			1	1/2
y_{11}				1

Table 3.6-8*Comparison Matrix* 8

f_4	y_{12}	y_{13}	y_{14}	y_{15}
y_{12}	1	1/3	2	1
y_{13}		1	4	2
y_{14}			1	1/3
y_{15}				1

$$C3 = \begin{bmatrix} 0.1411 \\ 0.3298 \\ 0.1411 \\ 0.3880 \end{bmatrix} \qquad \lambda_{max} = 4.1545 \qquad CI = 0.0515$$

$$C4 = \begin{bmatrix} 0.1933 \\ 0.4734 \\ 0.0966 \\ 0.2367 \end{bmatrix} \qquad \lambda_{max} = 4.0412 \qquad CI = 0.0137$$

Table 3.6-9 *Comparison Matrix* 9

f_5	y_{16}	y_{17}	y_{18}
y_{16}	1	1	3
y_{17}		1	3
y_{18}			1

$$C5 = \begin{bmatrix} 0.4286 \\ 0.4286 \\ 0.1429 \end{bmatrix} \qquad \lambda_{max} = 3.0000 \qquad CI = 0.0000$$

Level III contains the comparisons of vessel performance with respect to each attribute. To understand the comparisons a brief description of the vessels follows:

General cargo type (a_1):

Physical performance:

- Propulsion: single screw
- Speed and size: slightly smaller than a_2 and a_3
- Draught: more or less the same as in a_3 but higher than that in a_2.
- Handling equipment: simple crane with modest SWL
- Cargo package: almost any type of package except liquid with moderate safety of cargo stowage
- General features: more or less the same as a_2 and a_3 (eg. the same expected economic life)
- Reliance on port facilities: moderate
- Cargo handling time: the highest amongst the options

Cost performance:

- Capital cost: smaller than that for a_2 and a_3
- Port costs (Port Charges): high
- Operating cost: since it is the simplest type of vessel, it is assumed to have the lowest operating cost
- Unit cargo handling cost: high

Ro-ro type (a_2):

Physical performance:

- Propulsion: twin screw
- Speed and size: slightly bigger than a_1, but smaller than a_3
- Draught: the smallest amongst the options
- Handling equipment: none
- Cargo package: unitized cargo only ie. palletized, container etc.
- General features: more or less the same as a_1 and a_3 (eg. the same expected economic life)
- Reliance on port facilities: highly dependent
- Cargo handling time: the lowest cargo handling time

Cost performance:

- Capital cost: higher than that for a_1, but lower than that for a_3
- Port costs (Port Charges): moderate
- Operating cost: slightly higher than a_1, but lower than a_3
- Unit cargo handling cost: moderate, somewhere between that for a_1 and a_3

Full-container type (a_3):

Physical performance:

- Propulsion: single screw
- Speed and size: slightly greater carrying capacity than a_2, and greater than a_1
- Draught: more or less the same as in a_1
- Handling equipment: sophisticated cranes on board
- Cargo package: ISO container only
- General features: more or less the same as a_1 and a_2 (eg. the same expected economic life)

- Reliance on port facilities: moderate
- Cargo handling time: moderate, somewhere between that for a_1 and a_2

Cost performance:

- Capital cost: the highest capital cost
- Port costs (Port Charges): low
- Operating cost: high
- Unit cargo handling cost: relatively low

For the attribute y_1, the three vessel types are evaluated in a pairwise fashion as follows.

Table 3.6-10 *Comparison Matrix* 10

y_1	a_1	a_2	a_3
a_1	1	1/3	1/2
a_2		1	2
a_3			1

$$C6 = \begin{bmatrix} 0.1634 \\ 0.5396 \\ 0.2970 \end{bmatrix} \qquad \lambda_{max} = 3.0092 \qquad CI = 0.0046$$

For the attribute Roundtrip Time therefore, ro-ro cargo type is the best amongst the three types. Similarly,

Table 3.6-11 *Comparison Matrix* 11

y_2	a_1	a_2	a_3
a_1	1	1/3	1/2
a_2		1	2
a_3			1

Table 3.6-12 *Comparison Matrix* 12

y_3	a_1	a_2	a_3
a_1	1	1/2	2
a_2		1	3
a_3			1

$$C7 = \begin{bmatrix} 0.1634 \\ 0.5396 \\ 0.2970 \end{bmatrix} \qquad \lambda_{max} = 3.0092 \qquad CI = 0.0046$$

$$C8 = \begin{bmatrix} 0.2970 \\ 0.5396 \\ 0.1634 \end{bmatrix} \qquad \lambda_{max} = 3.0092 \qquad CI = 0.0046$$

Table 3.6-13 *Comparison Matrix* 13

y_4	a_1	a_2	a_3
a_1	1	2	3
a_2		1	2
a_3			1

Table 3.6-14 *Comparison Matrix* 14

y_5	a_1	a_2	a_3
a_1	1	1/2	1/3
a_2		1	1/3
a_3			1

$$C9 = \begin{bmatrix} 0.5396 \\ 0.2970 \\ 0.1634 \end{bmatrix} \qquad \lambda_{max} = 3.0092 \qquad CI = 0.0046$$

$$C10 = \begin{bmatrix} 0.1571 \\ 0.2493 \\ 0.5936 \end{bmatrix} \qquad \lambda_{max} = 3.0536 \qquad CI = 0.0268$$

Table 3.6-15 *Comparison Matrix* 15

y_6	a_1	a_2	a_3
a_1	1	2	3
a_2		1	2
a_3			1

Table 3.6-16 *Comparison Matrix* 16

y_7	a_1	a_2	a_3
a_1	1	1/2	1/3
a_2		1	1/3
a_3			1

$$C11 = \begin{bmatrix} 0.5396 \\ 0.2970 \\ 0.1634 \end{bmatrix} \qquad \lambda_{\max} = 3.0092 \qquad CI = 0.0046$$

$$C12 = \begin{bmatrix} 0.1571 \\ 0.2493 \\ 0.5936 \end{bmatrix} \qquad \lambda_{\max} = 3.0536 \qquad CI = 0.0268$$

Table 3.6-17 *Comparison Matrix* 17

y_8	a_1	a_2	a_3
a_1	1	2	3
a_2		1	2
a_3			1

Table 3.6-18 *Comparison Matrix* 18

y_9	a_1	a_2	a_3
a_1	1	1/2	1/3
a_2		1	2
a_3			1

$$C13 = \begin{bmatrix} 0.5396 \\ 0.2970 \\ 0.1634 \end{bmatrix} \qquad \lambda_{\max} = 3.0092 \qquad CI = 0.0046$$

$$C14 = \begin{bmatrix} 0.1677 \\ 0.4836 \\ 0.3487 \end{bmatrix} \qquad \lambda_{\max} = 3.1356 \qquad CI = 0.0678$$

Table 3.6-19 *Comparison Matrix* 19

y_{10}	a_1	a_2	a_3
a_1	1	2	3
a_2		1	2
a_3			1

Table 3.6-20 *Comparison Matrix* 20

y_{11}	a_1	a_2	a_3
a_1	1	1/2	1/3
a_2		1	1/2
a_3			1

$$C15 = \begin{bmatrix} 0.5396 \\ 0.2970 \\ 0.1634 \end{bmatrix} \qquad \lambda_{max} = 3.0092 \qquad CI = 0.0046$$

$$C16 = \begin{bmatrix} 0.1634 \\ 0.2970 \\ 0.5396 \end{bmatrix} \qquad \lambda_{max} = 3.0092 \qquad CI = 0.0046$$

Table 3.6-21 *Comparison Matrix* 21

y_{12}	a_1	a_2	a_3
a_1	1	1/2	1/3
a_2		1	1/2
a_3			1

Table 3.6-22 *Comparison Matrix* 22

y_{13}	a_1	a_2	a_3
a_1	1	2	3
a_2		1	1/2
a_3			1

$$C17 = \begin{bmatrix} 0.1634 \\ 0.2970 \\ 0.5396 \end{bmatrix} \qquad \lambda_{max} = 3.0092 \qquad CI = 0.0046$$

$$C18 = \begin{bmatrix} 0.5472 \\ 0.1897 \\ 0.2631 \end{bmatrix} \qquad \lambda_{max} = 3.1356 \qquad CI = 0.0678$$

Table 3.6-23 *Comparison Matrix* 23

y_{14}	a_1	a_2	a_3
a_1	1	1	1
a_2		1	1
a_3			1

Table 3.6-24 *Comparison Matrix* 24

y_{15}	a_1	a_2	a_3
a_1	1	1/2	1/3
a_2		1	1/2
a_3			1

$$C19 = \begin{bmatrix} 0.3333 \\ 0.3333 \\ 0.3333 \end{bmatrix} \qquad \lambda_{max} = 3.0000 \qquad CI = 0.0000$$

$$C20 = \begin{bmatrix} 0.1634 \\ 0.2970 \\ 0.5396 \end{bmatrix} \qquad \lambda_{max} = 3.0092 \qquad CI = 0.0046$$

Table 3.6-25 *Comparison Matrix* 25

y_{16}	a_1	a_2	a_3
a_1	1	1/3	1
a_2		1	3
a_3			1

Table 3.6-26 *Comparison Matrix* 26

y_{17}	a_1	a_2	a_3
a_1	1	2	1/3
a_2		1	1/3
a_3			1

$$C21 = \begin{bmatrix} 0.2000 \\ 0.6000 \\ 0.2000 \end{bmatrix} \qquad \lambda_{max} = 3.0000 \qquad CI = 0.0000$$

$$C22 = \begin{bmatrix} 0.2493 \\ 0.1571 \\ 0.5936 \end{bmatrix} \qquad \lambda_{max} = 3.0536 \qquad CI = 0.0268$$

Table 3.6-27 *Comparison Matrix 27*

y_{18}	a_1	a_2	a_3
a_1	1	3	2
a_2		1	1/2
a_3			1

$$C23 = \begin{bmatrix} 0.5396 \\ 0.1634 \\ 0.2970 \end{bmatrix} \qquad \lambda_{max} = 3.0092 \qquad CI = 0.0046$$

The weights can be combined in the following manner. For attribute f_1,

$$\begin{matrix} C6 & C7 & C8 & & C1 & & V1 \end{matrix}$$

$$\begin{bmatrix} 0.1634 & 0.1634 & 0.2970 \\ 0.5396 & 0.5396 & 0.5396 \\ 0.2970 & 0.2970 & 0.1634 \end{bmatrix} \times \begin{bmatrix} 0.4 \\ 0.4 \\ 0.2 \end{bmatrix} = \begin{bmatrix} 0.1901 \\ 0.5396 \\ 0.2703 \end{bmatrix}$$

Thus, on the basis of the quality of service, ro-ros are found to be the best for this example. For attribute f_2,

$$\begin{matrix} C9 & C10 & C11 & C12 & & C2 & & V2 \end{matrix}$$

$$\begin{bmatrix} 0.5396 & 0.1571 & 0.5396 & 0.1571 \\ 0.2970 & 0.2493 & 0.2970 & 0.2493 \\ 0.1634 & 0.5936 & 0.1634 & 0.5936 \end{bmatrix} \times \begin{bmatrix} 0.1225 \\ 0.3959 \\ 0.1142 \\ 0.3674 \end{bmatrix} = \begin{bmatrix} 0.2476 \\ 0.2606 \\ 0.4918 \end{bmatrix}$$

On the basis of the characteristics of the commodities on the route being considered, container ships appear to be the best. Similarly, for attribute f_3,

$$\begin{matrix} C13 & C14 & C15 & C16 \end{matrix} \qquad C3 \qquad V3$$

$$\begin{bmatrix} 0.5396 & 0.1677 & 0.5396 & 0.1634 \\ 0.2970 & 0.4836 & 0.2970 & 0.2970 \\ 0.1634 & 0.3487 & 0.1634 & 0.5396 \end{bmatrix} \times \begin{bmatrix} 0.1411 \\ 0.3298 \\ 0.1411 \\ 0.3880 \end{bmatrix} = \begin{bmatrix} 0.2710 \\ 0.3585 \\ 0.3705 \end{bmatrix}$$

For attribute f_4,

$$\begin{matrix} C17 & C18 & C19 & C20 \end{matrix} \qquad C4 \qquad V4$$

$$\begin{bmatrix} 0.1634 & 0.5472 & 0.3333 & 0.1634 \\ 0.2970 & 0.1897 & 0.3333 & 0.2970 \\ 0.5396 & 0.2631 & 0.3333 & 0.5396 \end{bmatrix} \times \begin{bmatrix} 0.1933 \\ 0.4734 \\ 0.0966 \\ 0.2367 \end{bmatrix} = \begin{bmatrix} 0.3615 \\ 0.2497 \\ 0.3888 \end{bmatrix}$$

For attribute f_5,

$$\begin{matrix} C21 & C22 & C23 \end{matrix} \qquad C5 \qquad V5$$

$$\begin{bmatrix} 0.2000 & 0.2493 & 0.5396 \\ 0.6000 & 0.1571 & 0.1634 \\ 0.2000 & 0.5936 & 0.2970 \end{bmatrix} \times \begin{bmatrix} 0.4286 \\ 0.4286 \\ 0.1429 \end{bmatrix} = \begin{bmatrix} 0.2697 \\ 0.3478 \\ 0.3826 \end{bmatrix}$$

The composite priority for all factors with respect to all the decision makers can be found in a similar way

$$\begin{matrix} W2 & W3 & W4 \end{matrix} \qquad W1 \qquad X1$$

$$\begin{bmatrix} 0.1497 & 0.2253 & 0.3664 \\ 0.1698 & 0.3935 & 0.1556 \\ 0.2572 & 0.1852 & 0.0889 \\ 0.3279 & 0.0840 & 0.0889 \\ 0.0954 & 0.1120 & 0.3002 \end{bmatrix} \times \begin{bmatrix} 0.3333 \\ 0.3333 \\ 0.3333 \end{bmatrix} = \begin{bmatrix} 0.2472 \\ 0.2396 \\ 0.1771 \\ 0.1669 \\ 0.1692 \end{bmatrix} \begin{matrix} (1) \\ (2) \\ (3) \\ (5) \\ (4) \end{matrix}$$

From $X1$, it will be seen that the most important factor for all the decision makers in this instance is the quality of service (Factor f_1) and the order of importance is f_1, f_2, f_3, f_5 and f_4.

The global weights for the given vessels can now be found as follows

$$\begin{array}{c} \\ O1 \\ O2 \\ O3 \end{array} \begin{array}{c} \begin{array}{ccccc} V1 & V2 & V3 & V4 & V5 \end{array} \\ \begin{bmatrix} 0.1901 & 0.2476 & 0.2710 & 0.3615 & 0.2697 \\ 0.5396 & 0.2606 & 0.3585 & 0.2497 & 0.3478 \\ 0.2703 & 0.4918 & 0.3705 & 0.3888 & 0.3826 \end{bmatrix} \end{array} \times \overset{X1}{\begin{bmatrix} 0.2472 \\ 0.2396 \\ 0.1771 \\ 0.1669 \\ 0.1692 \end{bmatrix}} = \overset{P1}{\begin{bmatrix} 0.2603 \\ 0.3598 \\ 0.3799 \end{bmatrix}}$$

The final weights for selection of the vessel type is

$$P1 = \begin{array}{c} O1 \\ O2 \\ O3 \end{array} \begin{bmatrix} 0.2603 \\ 0.3598 \\ 0.3799 \end{bmatrix} \quad \begin{array}{c} (3) \\ (2) \\ (1) \end{array}$$

Thus, full-container vessels are the best with ro-ros a close runner-up.

The influence of changes in weights at a given level may be seen in the following. Assuming the shipper is strongly more important than other two decision makers (ie. the Shipping Company and the Government) with scale 6, then the matrix scheme becomes

Table 3.6-28 *Comparison Matrix* 28

vessel choice	d_1	d_2	d_3
d_1	1	1/6	1
d_2		1	6
d_3			1

$$W1 = \begin{bmatrix} 0.125 \\ 0.725 \\ 0.150 \end{bmatrix} \qquad \lambda_{max} = 3.000 \qquad CI = 0.000$$

Then vector $X1$ (for importance of the factors) becomes

$$X1 = \begin{bmatrix} 0.3442 \\ 0.2628 \\ 0.1638 \\ 0.1089 \\ 0.1203 \end{bmatrix} \begin{matrix} (1) \\ (2) \\ (3) \\ (5) \\ (4) \end{matrix}$$

and it can be seen that the order of importance for factors is unchanged.

The new priority may be seen as follows

$$\begin{matrix} & V1 & V2 & V3 & V4 & V5 \\ O1 & 0.1901 & 0.2476 & 0.2710 & 0.3615 & 0.2697 \\ O2 & 0.5396 & 0.2606 & 0.3585 & 0.2497 & 0.3478 \\ O3 & 0.2703 & 0.4918 & 0.3705 & 0.3888 & 0.3826 \end{matrix} \times \overset{X1}{\begin{bmatrix} 0.3442 \\ 0.2628 \\ 0.1638 \\ 0.1089 \\ 0.1203 \end{bmatrix}} = \overset{P1}{\begin{bmatrix} 0.2467 \\ 0.3820 \\ 0.3713 \end{bmatrix}}$$

The final result for priority of the vessel type is given by

$$P1 = \begin{matrix} O1 \\ O2 \\ O3 \end{matrix} \begin{bmatrix} 0.2467 \\ 0.3820 \\ 0.3713 \end{bmatrix} \begin{matrix} (3) \\ (1) \\ (2) \end{matrix}$$

Now, ro-ros look marginally better than container vessels.

Obviously, the conclusions are dependent on the weights but the method is sensitive to the changes in judgements and permits non-quantifiable features to be taken into account within the decision making process.

3.3.2 UTA Method and Application

The Additive UTility (UTA) function method is proposed by E. Jacquet-Lagreze and J. Siskos [Jacquet-Lagreze and Siskos 1982]. The UTA method provides an alternative way for eliciting preference information in terms of the subjective ranking of a subset of alternative designs. UTA then uses such information to assess a set of additive utility functions based on the multicriteria evaluations of these designs. Linear programming is used to estimate the parameters of utility functions. Sensitivity analysis is also conducted to generate a mean utility function by means of post-optimality analysis.

The computational steps of UTA is summarized as follows.

Step 1: Select a subset of designs and rank these designs subjectively with regard to all attributes. Construct a decision matrix for these designs as shown in Table 3.1

Step 2: Define an additive utility function as

$$U(y) = \sum_{i=1}^{k} u_i(y_i) \tag{3.43}$$

where $u_i(y_i)$ is the marginal utility function for the attribute y_i. u_i is assumed to be a piecewise linear function. Let $[y_i^- \; y_i^*]$ be the intervals in which the values of the attribute y_i are defined. Cut the interval $[y_i^- \; y_i^*]$ into $(\alpha_i - 1)$ equal intervals $[y_i^j \; y_i^{j+1}]$ $(j=1, \cdots, \alpha_i-1)$ where $y_i^1 = y_i^-$ and $y_i^{\alpha_i} = y_i^*$. The end points y_i^j of these equal intervals are then given by

$$y_i^j = y_i^- + \frac{j-1}{\alpha_i - 1}(y_i^* - y_i^-) \qquad j=1, \cdots, \alpha_i-1 \tag{3.44}$$

The marginal utility of a design a_l is approximated by a linear interpolation. Let $u_i(y_i^j)$ be the utility of y_i^j. If $y_i(a_l) \in [y_i^j \; y_i^{j+1}]$, $u_i(y_i(a_l))$ is then estimated by

$$\hat{u}_i[y_i(a_l)] = u_i(y_i^j) + \frac{y_i(a_l) - y_i^j}{y_i^{j+1} - y_i^j}[u_i(y_i^{j+1}) - u_i(y_i^j)] \tag{3.45}$$

where $u_i(y_i^j)$ for $j = 1, \cdots, \alpha_i-1$ are parameters to be estimated. The utility of a design a_l is then given by

$$u[y(a_l)] = \sum_{i=1}^{k} u_i[y_i(a_l)] = \sum_{i=1}^{k} \hat{u}_i[y_i(a_l)] + \sigma(a_l) \tag{3.46}$$

where $\sigma(a_l)$ is estimation error.

Step 3: Acquire preference relations. Let P stand for strict preference relation and I for indifference relation. Then the following properties generally hold for a utility function u

$$u[y(a_l)] > u[y(a_h)] \Leftrightarrow a_l P a_h \tag{3.47}$$

$$u[y(a_l)] = u[y(a_h)] \Leftrightarrow a_l I a_h \tag{3.48}$$

or

$$\sum_{i=1}^{k} \hat{u}_i[y_i(a_l)] - \hat{u}_i[y_i(a_h)] + \sigma(a_l) - \sigma(a_h) \geq \delta \Leftrightarrow a_l P a_h \tag{3.49}$$

$$\sum_{i=1}^{k} \hat{u}_i[y_i(a_l)] - \hat{u}_i[y_i(a_h)] + \sigma(a_l) - \sigma(a_h) = 0 \Leftrightarrow a_l I a_h \tag{3.50}$$

where δ is a small positive real number which should be chosen so that a_l and a_h can be separated significantly.

Step 4: Assume that all marginal utility functions are monotonic, either non-increasing or non-decreasing. A non-increasing marginal utility function $u_i(y_i)$ must satisfy the following conditions

$$u_i(y_i^{j+1}) - u_i(y_i^j) \leq 0 \quad j=1, 2, ..., \alpha_i-1 \tag{3.51}$$

A non-decreasing $u_i(y_i)$ must satisfy

$$u_i(y_i^{j+1}) - u_i(y_i^j) \geq 0 \quad j=1, 2, ..., \alpha_i-1 \tag{3.52}$$

Step 5: Construct the following linear programme to estimate the utilities $u_i(y_i^j)$, $j=1, 2, ..., \alpha_i-1$; $i=1, ..., k$

$$\begin{aligned}
\min \quad & F = \sum_{l=1}^{n} \sigma(a_l) \\
s.t. \quad & \sum_{i=1}^{k} \{\hat{u}_i[y_i(a_l)] - \hat{u}_i[y_i(a_h)]\} + \sigma(a_l) - \sigma(_h) \geq \delta \qquad \text{if } a_l P a_h \\
& \sum_{i=1}^{k} \{\hat{u}_i[y_i(a_l)] - \hat{u}_i[y_i(a_h)]\} + \sigma(a_l) - \sigma(_h) = 0 \qquad \text{if } a_l I a_h \\
& u_i(y_i^{j+1}) - u_i(y_i^j) \geq 0 \quad \text{for } all\ j, \quad \text{if } u_i \ is\ non-decreasing \\
& u_i(y_i^j) - u_i(y_i^{j+1}) \geq 0 \quad \text{for } all\ j, \quad \text{if } u_i \ is\ non-increasing \\
& \sum_{i=1}^{k} u_i(y_i^*) = 1 \\
& u_i(y_i^-)=0,\ u_i(y_i^j) \geq 0,\ \sigma(a) \geq 0;\ j=1,2, \cdots ,\alpha_i-1,\ i=1, \cdots ,k
\end{aligned} \tag{3.53}$$

where F is a linear objective function for minimization of the total deviation. Solving the linear programming problem (3.53), we can obtain the optimal estimation of the marginal utilities, $\hat{u}_i(y_i^j)$, $j=1, \cdots, \alpha_i-1$; $i=1, \cdots ,k$.

Step 6: Generate the mean utilities. Generally problem (3.53) may have multiple optimal solutions. It is then necessary to explore other optimal or sub-optimal solutions around the generated optimum. Let F^* be the optimal value of the objective function of (3.53). Suppose $k(F^*)$ is a real non-negative threshold which is zero or a very small proportion of F^*. Solve the following linear programming problems

$$\min \quad \sum_{i=1}^{k} \rho_i u_i(y_i^*)$$

$$s.t. \quad \textit{the constraint set of } (3.53) \text{ and } \sum_{l=1}^{k} \sigma(a_l) \le F^* + k(F^*) \tag{3.54}$$

$$\max \quad \sum_{i=1}^{k} \rho_i u_i(y_i^*)$$

$$s.t. \quad \textit{the constraint set of } (3.53) \text{ and } \sum_{l=1}^{k} \sigma(a_l) \le F^* + k(F^*) \tag{3.55}$$

with $\rho_i = 1$ or 0 for all i. Let $\underline{u}_i(y_i^j)$ and $\overline{u}_i(y_i^j)$ be the solutions of problems (3.54) and (3.55), respectively. Then, the mean marginal utility of y_i^j, denoted by $\hat{u}_i(y_i^j)$, is given by

$$\hat{\overline{u}}_i(y_i^j) = \frac{1}{2k} \sum_{j=1}^{k} (\underline{u}_i(y_i^j) + \overline{u}_i(y_i^j)) \tag{3.56}$$

Step 7: Use the assessed utility functions to rank all designs. A better design should have a large value of utility.

The UTA method provides an alternative way to elicit and represent preferences. However, this method also assumes that attributes are preferentially independent as it adopts an additive utility function.

The following example was initially reported in reference [Jacquet-Lagreze and Siskos 1982]. The problem is to rank a set of twenty eight cars based on six performance attributes. The decision maker knew ten of the cars and provided a subjective ranking of the ten cars. The evaluation data and the subjective ranking of these ten cars are as shown in Table 3.7.

Table 3.7 is then used to estimate utility functions. In order to assess utility functions using the UTA method, the attribute variation intervals $[y_i^- \;\; y_i^*]$ and the values of the parameters α_i and δ have to be determined. The value chosen for δ is 0.01 and the other values are given by Table 3.8.

Table 3.7 *A Ranking of Ten Reference Cars*

Reference cars	subjective ranking	Multiattribute evaluations					
		Maximal speed (km) (y_1)	Consumption in town (lt/100 km) (y_2)	Consumption at 120 km/h (lt/100 km) (y_3)	Horse power (CV) (y_4)	Space (m^2) (y_5)	Price (francs) (y_6)
Car 1 (C_1)	1	173	11.4	10.01	10	7.88	49500
Car 2 (C_2)	2	176	12.3	10.48	11	7.96	46700
Car 3 (C_3)	3	142	8.2	7.30	5	5.65	32100
Car 4 (C_4)	4	148	10.5	9.61	7	6.15	39150
Car 5 (C_5)	5	178	14.5	11.05	13	8.06	64700
Car 6 (C_6)	6	180	13.6	10.40	13	8.47	75700
Car 7 (C_7)	7	182	12.7	12.26	11	7.81	68593
Car 8 (C_8)	8	145	14.3	12.95	11	8.38	55000
Car 9 C_9)	9	161	8.6	8.42	7	5.11	35200
Car 10 (C_{10})	10	117	7.2	6.75	3	5.81	24800

Table 3.8 *Chosen Values for the Parameters*

Attributes	Parameters		
	y_i^-	y_i^*	α_i
y_1	110	190	5
y_2	15	7	4
y_3	13	6	4
y_4	3	13	5
y_5	5	9	4
y_6	80000	20000	5

From the information given by Tables 3.7 and 3.8, the linear programming problem (3.53) can be constructed, in which there are 31 constraints and 32 variables. Solving problem (3.53), we have F^*=0.0, which means that the estimated utility is perfectly consistent with the subjective ranking and the multicriteria evaluations given in Table 3.7. The obtained marginal utilities $\hat{u}_i(y_i^j)$ at the designated end points y_i^j are as shown by Table 3.9.

To generate mean utility functions, let $k(F^*) = 0.009$ (just smaller than δ). Thus, twelve linear programming problems as defined by equations (3.54) and (3.55) for i=1, $\cdots$,6 are then solved. The mean marginal utilities are calculated by equation (3.56) and shown by $\hat{\bar{u}}_i(y_i^j)$ as in Tables 3.9.

Table 3.9-1 *Estimated Marginal Utilities*

End points (j)	Maximal speed (km)			Consumption in town (lt/100 km)		
	y_1^j	$\hat{u}_1(y_1^j)$	$\bar{u}_1(y_1^j)$	y_2^j	$\hat{u}_2(y_2^j)$	$\bar{u}_2(y_2^j)$
1	110	0.000	0.000	7.0	0.041	0.104
2	130	0.000	0.090	9.7	0.041	0.099
3	150	0.176	0.135	12.7	0.000	0.014
4	170	0.176	0.137	15.7	0.000	0.000
5	190	0.176	0.145			

Table 3.9-2 *Estimated Marginal Utilities*

End points (j)	Consumption at 120 km/h (lt/100 km)			Horse power (CV)		
	y_3^j	$\hat{u}_3(y_3^j)$	$\bar{u}_3(y_3^j)$	y_4^j	$\hat{u}_4(y_4^j)$	$\bar{u}_4(y_4^j)$
1	6.0	0.169	0.170	3.0	0.000	0.000
2	8.3	0.000	0.038	5.5	0.171	0.185
3	10.6	0.000	0.032	8.0	0.171	0.203
4	13.0	0.000	0.032	10.5	0.171	0.205
5				13.0	0.191	0.214

Table 3.9-3 *Estimated Marginal Utilities*

End points (j)	Space (m^2)			Price (francs)		
	y_5^j	$\hat{u}_5(y_5^j)$	$\bar{u}_5(y_5^j)$	y_5^j	$\hat{u}_5(y_5^j)$	$\bar{u}_5(y_5^j)$
1	5.0	0.000	0.000	20000	0.306	0.188
2	6.3	0.103	0.120	35000	0.059	0.066
3	7.7	0.115	0.174	50000	0.059	0.053
4	9.0	0.115	0.179	65000	0.013	0.028
5				80000	0.000	0.000

The utility of a car may then be calculated using equations (3.45) and (3.46) where $u_i(y_i^j)$ is replaced by $\hat{u}_i(y_i^j)$ or $\bar{u}_i(y_i^j)$ as given in Tables 3.9. In this way, the utilities of the ten reference cars are obtained as follows

$$[u(C_1)\ u(C_2)\ u(C_3)\ u(C_4)\ u(C_5)\ u(C_6)\ u(C_7)\ u(C_8)\ u(C_9)\ u(C_{10})]$$
$$= [0.536\ 0.526\ 0.516\ 0.506\ 0.496\ 0.486\ 0.476\ 0.466\ 0.456\ 0.446] \tag{3.57}$$

Thus the ranking of the ten cars obtained on the basis of the magnitude of their utilities is the same as that given by Table 3.7.

3.3.3 TOPSIS Method and Application

Hwang and Yoon developed the Technique for Order Preference by Similarity to Ideal Solution (TOPSIS), based upon the concept that the chosen alternative should have the shortest distance to the ideal point and the furthest distance from the negative ideal point [Hwang and Yoon 1981]. Similar to the UTA methods, TOPSIS also requires a decision matrix as input evaluation data but uses given relative weights as the representation of preference information. In addition to its simplicity, the TOPSIS method only assumes that each attribute takes either monotonically increasing or monotonically decreasing utility. For most MADM problems, this assumption is acceptable.

The computational steps of TOPSIS can be summarized as follows.

Step 1: Suppose a MADM problem can be represented as in Table 3.1. Then construct the normalized decision matrix whose elements are defined by

$$z_{ij} = \frac{y_{ij}}{\sqrt{\sum_{i=1}^{n} y_{ij}^2}}, \quad i=1,\cdots,n;\ j=1,\cdots,k \tag{3.58}$$

Consequently, each attribute has the same unit length of vector.

Step 2: Formulate the weighted normalized decision matrix whose elements are given by

$$x_{ij} = w_j z_{ij} \quad i=1,\cdots,n;\ j=1,\cdots,k \tag{3.59}$$

where w_j is the weight of the jth attribute.

Step 3: Define the ideal point a^* and the negative ideal point (nadir) a^- as follows

$$a^* = \{(\max_i x_{ij} \mid j\in J), (\min_i x_{ij} \mid j\in \hat{J}) \mid i=1,\cdots,n\} \tag{3.60}$$

$$= \{x_1^*, x_2^*, \dots, x_k^*\}$$

$$a^- = \{(\min_i x_{ij} \mid j\in J), (\max_i x_{ij} \mid j\in \hat{J}) \mid i=1,\cdots,n\} \tag{3.61}$$

$$= \{x_1^-, x_2^-, ..., x_k^-\}$$

where J is the index set of benefit attributes and $\hat{J}$ the index set of cost attributes.

Step 4: Calculate the distance of a design to the ideal point,

$$S_i^* = \sqrt{\sum_{j=1}^{k}(x_{ij} - x_j^*)^2} \quad i=1, \cdots, n \tag{3.62}$$

and the distance of the design from the negative ideal point,

$$S_i^- = \sqrt{\sum_{j=1}^{k}(x_{ij} - x_j^-)^2} \quad i=1, \cdots, n \tag{3.63}$$

Step 5: Calculate the relative closeness of each design to the ideal point

$$C_i^* = S_i^- / (S_i^- + S_i^*) \quad 0 \leq C_i^* \leq 1, \quad i=1, \cdots, n \tag{3.64}$$

Step 6: Rank the alternatives based on the magnitude of closeness C_i^*. If $C_i^* > C_j^*$, then a_i is preferred to a_j.

The TOPSIS method can produce a clear preference order of a set of competing designs. However, TOPSIS suffers from two main weak points. First of all, the definition of the separation between each alternative and the ideal point or the negative ideal point, measured by the n-dimensional Euclidean distance in the attribute space defined by the weighted normalized decision matrix, is rather sensitive to weights. These weights may only be subjectively evaluated and hence are often inaccurate. The inaccuracy may become worse with the increase of the number of attributes. Secondly, direct and unlimited compensation among all attributes is assumed in the definition of distance. In a MADM problem, however, some attributes may not be allowed to compensate for each other in such a simple way. Such compensation may ignore important features of a good design with respect to some attributes and consequently the design may be unexpectedly dominated by another design with better average features with regard to all attributes.

To demonstrate the application of the TOPSIS method, a fighter aircraft selection problem is taken for example, as shown by Table 3.10 [Huang and Yoon 1981]. In Table 3.10, a^* and a^- are two dummy alternatives used as reference so that all aeroplanes can be properly evaluated, as they represent the notional "best" and "worst" aircraft respectively.

Table 3.10 *A Fighter Aircraft Selection Problem*

Alternative aeroplanes (a_i)	Attributes y_j					
	Maximum speed (*Mach*) y_1	Ferry range ($NM \times 10^3$) y_2	Maximum pay load ($pounds \times 10^3$) y_3	Acquisition cost ($pounds \times 10^6$) y_4	Reliability (*high–low*) y_5	Manoeuvrability (*high–low*) y_6
a_1	2.0	1.5	2.0	5.5	average	very high
a_2	2.5	2.7	1.8	6.5	low	average
a_3	1.8	2.0	2.1	4.5	high	high
a_4	2.2	1.8	2.0	5.0	average	average
a^*	2.8	3.0	2.5	4.0	extremely high	extremely high
a^-	1.5	1.0	1.5	7.0	extremely low	extremely low

The qualitative attributes y_5 and y_6 were quantified using an interval scale. The decision matrix is quantified as follows, where the minus sign "-" in the fourth column means that acquisition cost is for minimization.

$$D = \begin{bmatrix} 2.0 & 1.5 & 2.0 & -5.5 & 5.0 & 9.0 \\ 2.5 & 2.7 & 1.8 & -6.5 & 3.0 & 5.0 \\ 1.8 & 2.0 & 2.1 & -4.5 & 7.0 & 7.0 \\ 2.2 & 1.8 & 2.0 & -5.0 & 5.0 & 5.0 \\ 2.8 & 3.0 & 2.5 & -4.0 & 10.0 & 10.0 \\ 1.5 & 1.0 & 1.5 & -7.0 & 0.0 & 0.0 \end{bmatrix} \tag{3.65}$$

The relative weights $\hat{W}$ of the six attributes are given by the following scores (see Sections 3.2.1 and 3.2.2) and normalized by W

$$\hat{W} = [9 \quad 2 \quad 2 \quad 3 \quad 3 \quad 9]^T \tag{3.66}$$

$$W = [0.3214 \quad 0.0714 \quad 0.0714 \quad 0.1071 \quad 0.1071 \quad 0.3214]^T \tag{3.67}$$

Following the calculation steps of TOPSIS, the six options can be ranked as follows. First, normalize the decision matrix (3.65) using equations (3.58). We thus obtain the following normalized decision matrix

$$\bar{D} = \begin{bmatrix} 0.3752 & 0.2899 & 0.4070 & -0.4068 & 0.3467 & 0.5379 \\ 0.4690 & 0.5217 & 0.3663 & -0.4808 & 0.2080 & 0.2988 \\ 0.3376 & 0.3865 & 0.4273 & -0.3329 & 0.4854 & 0.4183 \\ 0.4127 & 0.3478 & 0.4070 & -0.3699 & 0.3467 & 0.2988 \\ 0.5252 & 0.5797 & 0.5087 & -0.2959 & 0.6934 & 0.5976 \\ 0.2814 & 0.1932 & 0.3052 & -0.5178 & 0.0000 & 0.0000 \end{bmatrix} \quad (3.68)$$

Then use formula (3.59) to combine the given weights with the normalized multiattribute evaluations of the aeroplanes as given by equation (3.68)

$$\tilde{D} = \begin{bmatrix} 0.1206 & 0.0207 & 0.0291 & -0.0436 & 0.0371 & 0.1729 \\ 0.1507 & 0.0373 & 0.0262 & -0.0515 & 0.0223 & 0.0960 \\ 0.1085 & 0.0276 & 0.0305 & -0.0357 & 0.0520 & 0.1345 \\ 0.1326 & 0.0248 & 0.0291 & -0.0396 & 0.0371 & 0.0960 \\ 0.1688 & 0.0414 & 0.0363 & -0.0317 & 0.0743 & 0.1921 \\ 0.0904 & 0.0138 & 0.0218 & -0.0555 & 0.0000 & 0.0000 \end{bmatrix} \quad (3.69)$$

Obviously, the ideal and the negative ideal points are given by the last two rows of the matrix (3.69), that is

$$a^* = [0.1688 \quad 0.0414 \quad 0.0363 \quad -0.0317 \quad 0.0743 \quad 0.1921]^T \quad (3.70)$$

$$a^- = [0.0904 \quad 0.0138 \quad 0.0218 \quad -0.0555 \quad 0.0000 \quad 0.0000]^T \quad (3.71)$$

The relative closeness of each aeroplane to the ideal (dummy) alternative is calculated using equations (3.62) to (3.64) as follows

$$[u(a_1) \quad u(a_2) \quad u(a_3) \quad u(a_4) \quad u(a_5) \quad u(a_6)]$$

$$= [0.7243 \quad 0.5110 \quad 0.6272 \quad 0.5051 \quad 1.0000 \quad 0.0000] \quad (3.72)$$

the aeroplanes are therefore ranked as follows

$$a_5 \succ a_1 \succ a_3 \succ a_2 \succ a_4 \succ a_6 \quad (3.73)$$

a^* and a^- are two dummy aerolanes, taking the best and the worst values of the six attributes, respectively. They are of course ranked to be the best and the worst options, respectively. a_1 is ranked to be the best feasible option as it has "very high" maneuverability, which is very important, and also has good achievement levels on other attributes.

Now, let's examine the selection problem of the 13 efficient designs for the semi-submersible referred to in Section 3.2.4.3. The decision matrix for the problems is as shown by Table 3.3 and the weights of the 6 attributes (objectives) are given by equation (3.37). TOPSIS produces the following closeness indices of the designs

$$[u(a_1)\ u(a_2)\ u(a_3)\ u(a_4)\ u(a_5)\ u(a_6)$$
$$u(a_7)\ u(a_8)\ u(a_9)\ u(a_{10})\ u(a_{11})\ u(a_{12})\ u(a_{13})]$$
$$= [0.6688\quad 0.3306\quad 0.2232\quad 0.2847\quad 0.3020\quad 0.7266\quad 0.6682$$
$$0.6909\quad 0.6892\quad 0.5609\quad 0.5885\quad 0.5981\quad 0.6165]$$

which results in the following ranking of these designs

$$a_6 \succ a_8 \succ a_9 \succ a_1 \succ a_7 \succ a_{13} \succ a_{12} \succ a_{11} \succ a_{10} \succ a_2 \succ a_5 \succ a_4 \succ a_3$$

Note that the above ranking is vertually the same as that obtained by using the attribute y_6 solely. a_6 is ranked as the best design by TOPSIS as it is the cheapest design. However, a_6 is worse than other designs except a_1 in terms of the first five attributes.

3.3.4 CODASID Method and Applications

In the above three methods, it is always assumed that any attribute can be offset by others without any limit. In certain decision situations, however, compensation between attributes may only allowed to happen within some limit. This subsection introduces the CODASID method which can perform limited compensation analysis [Yang, Sen et al. 1997].

The CODASID method attempts to generate a clear preference order for alternative designs through COncordance and Discordance Analyses by Similarity to Ideal Designs (simply CODASID). It is based on an extended concordance analysis and a modified discordance analysis using raw data represented by a decision matrix, relative weights and veto threshold values on attributes. The original concordance analysis of ELECTRE is extended in order to make full use of the input data. The discordance analysis is modified to take into account limited compensation by using veto values on attributes.

The new concordance and discordance analyses are used to generate three new indices, namely a preference concordance index, an evaluation concordance index and a discordance index. These three indices provide independent measures for evaluation of each alternative design and span a new space for ultimate ranking of alternative designs. A linear goal programming model and a regulation procedure of trade-off weights of the three new indices are designed to account for the limited compensation in the new method. A distance measure is defined in the new space to capture the similarities between a feasible design and given reference designs, which may be, for example, the best/least preferred (or ideal/nadir) designs. The basic idea of defining such a distance measure originates from the TOPSIS method. The new distance measure, however, is more general and able to take into account the limited compensation referred to earlier.

The new method is therefore characterized by its capacity to handle the limited compensation, to allow the full utilization of raw data and to provide a systematic computational procedure. Such features may be desirable in certain decision situations such as at the preliminary design stages of large engineering products where a large number of candidate designs may be generated and need to be comparatively assessed.

3.3.4.1 *Information Requirement* and *Normalization*

A decision matrix defined by Table 3.1 contains information for multiattribute evaluations. Without loss of generality all attributes in Table 3.1 are assumed to be for maximization. More complex MADM problems may eventually be transformed to a generalized decision matrix [Yang and Singh 1994a][Yang and Sen 1994b]. Suppose relative weights of attributes are given by the following vector

$$W = [w_1 \ w_2 \ \cdots \ w_n]^T; \ w_j \geq 0, \text{ and } \sum_{j=1}^{n} w_j = 1 \tag{3.74}$$

where w_j is a normalized relative weight of attribute y_j.

The decision maker (designer) may also assign a veto threshold value to an attribute so that if the absolute difference in the values of two alternatives on the attribute is larger than the threshold value the alternative with the lower attribute value should never outrank the other, regardless of other attributes. Let vt_j represent the threshold value of attribute j. Then, the set of veto threshold values is given by

$$VT = [vt_1 \cdots vt_j \cdots vt_n]^T \tag{3.75}$$

Note that information involved in *VT* should not be inconsistent with that contained in Table 3.1 or that in *W*. Such inconsistency can be defined, identified and eliminated, if necessary, as discussed later in following sections.

Since the attributes are generally incommensurate, the decision matrix needs to be normalized so as to transform the various attribute scales into comparable scales. The following linear transformation is employed

$$r_{ij} = \frac{y_{ij} - y_j^{\min}}{y_j^{\max} - y_j^{\min}} \tag{3.76}$$

where $y_j^{\max}$ and $y_j^{\min}$ must be the best and the worst values of y_j. If they are already listed in the decision matrix (Table 3.1), then they are given by

$$y_j^{\max} = \max\{y_{1j} \quad y_{2j} \quad \cdots \quad y_{mj}\} \tag{3.77}$$

$$y_j^{\min} = \min\{y_{1j} \quad y_{2j} \quad \cdots \quad y_{mj}\} \tag{3.78}$$

Otherwise, $y_j^{\max}$ and $y_j^{\min}$ must be assigned by the decision maker. However, an assigned $y_j^{\max}$ must not be worse than the best of y_j as listed in Table 3.1 and $y_j^{\min}$ must not be better than the worst of y_j. It is suggested that two dummy designs be created to act as reference designs which take the best preferred value and the least preferred value of every attribute, respectively. These may then be added to the designs in Table 3.1. This is what was done for the aircraft selection problem of the previous section.

This process transforms all the attributes into the unique closed interval [0 1] with the best values of all the attributes corresponding to 1 and the worst corresponding to 0. In addition, $(r_{kj} - r_{lj})$ is proportional to $(y_{kj} - y_{lj})$, that is

$$r_{kj} - r_{lj} = \frac{1}{y_j^{\max} - y_j^{\min}}(y_{kj} - y_{lj}) \tag{3.79}$$

So, the difference between r_{kj} and r_{lj} proportionally reflects that between y_{kj} and y_{lj}. The decision matrix that has been normalized using equations (3.76) to (3.78) is represented by R

$$R = \begin{bmatrix} r_{11} & r_{12} & \cdots & r_{1n} \\ r_{21} & r_{22} & \cdots & r_{2n} \\ \cdots & \cdots & \cdots & \cdots \\ r_{m1} & r_{m2} & \cdots & r_{mn} \end{bmatrix} \tag{3.80}$$

The decision matrix and the relative weights are then combined by constructing the following weighted normalized decision matrix Z

$$Z = \begin{bmatrix} z_{11} & z_{12} & \cdots & z_{1n} \\ z_{21} & z_{22} & \cdots & z_{2n} \\ \cdots & \cdots & \cdots & \cdots \\ z_{m1} & z_{m2} & \cdots & z_{mn} \end{bmatrix} \tag{3.81}$$

where $z_{ij} = w_j r_{ij}$ $i=1, \cdots, m;\ j=1, \cdots, n.$

3.3.4.2 New Concordance and Discordance Analyses

All feasible alternatives as listed in Table 3.1 (exclusive of the two dummy designs if provided) are generally assumed to be nondominated alternatives as dominated ones can be readily deleted from further consideration. In other words, no single feasible alternative of Table 3.1 is absolutely better or worse than any other feasible ones in terms of all the attributes. The purpose of design evaluation, however, is to ultimately rank these alternative designs based on the raw data represented by the weights and the multiattribute evaluations. If one design a_k is assumed to be better than another a_l, evidence must be given as to why this is or is not the case.

Obviously, evidence, either supporting or refuting the above assumption, is always available from the raw data as both a_k and a_l are nondominated designs. An unbiased evaluation should thus take into account all pieces of evidence available. Alternative a_k may eventually be preferred to a_l only if there is more evidence supporting the assumption that a_k outranks a_l than that opposed to the assumption. The basic idea of the new concordance and discordance analyses is to facilitate this gathering and aggregating of evidence which either supports or opposes an assumption that one alternative outranks others. The analyses are also designed to treat limited compensation among attributes.

a) Concordance and discordance sets

Based on the decision matrix, all alternatives can be compared in a pairwise fashion with respect to each attribute. If $y_{kj} \geq y_{lj}$, for example, a_k is then preferred to a_l as far as the attribute y_j is concerned. In other words, if a_k is assumed to be preferred to a_l, denoted by $a_k P a_l$, and if $y_{kj} \geq y_{lj}$, the *jth* attribute y_j is then concordant with or supports the assumption; otherwise, y_j is discordant with or objects to the assumption. In this way, a concordance set and a discordance set for each pair of alternatives a_k and a_l can be defined as follows [Hwang and Yoon 1981][Yang, Sen et al. 1997].

$$C_{kl} = \{j \mid y_{kj} \geq y_{lj}, j=1, \cdots, n\}, \quad D_{kl} = \{j \mid y_{kj} < y_{lj}, j=1, \cdots, n\} \tag{3.82}$$

where $C_{kl} \cup D_{kl} = J = \{1 \ 2 \ \cdots \ n\}$.

Based on the decision matrix, the threshold values VT can be used to generate a set of non-outranking relationships between certain alternatives. As suggested by (3.75), if the following inequality is true

$$y_j(a_l) - y_j(a_k) \geq vt_j \quad \text{for } any \ j \in J \tag{3.83}$$

then the alternative a_k should never outrank the alternative a_l regardless of other attributes.

Thus, vt_j provides hard evidence objecting to the assumption that a_k outranks a_l. Such evidence is decisive and must not be compensated by any other evidence supporting the assumption. In other words, vt_j ($j=1, \cdots, n$) are used to set limits, within which compensation among attributes is permitted but beyond which any compensation is prohibited. A set of such non-outranking relationships, denoted by N_a, can be generated from Table 3.1 and VT as follows

$$N_a = \{(a_k, a_l) \mid y_j(a_l) - y_j(a_k) \geq vt_j \quad \text{for } any \ j \in J; \ k, l \in M; \ k \neq l\} \tag{3.84}$$

where $M = \{1 \ 2 \cdots m\}$.

If the information contained in Table 3.1, W and VT is consistent, the non-outranking relationships in N_a must not be in conflict with one another. Otherwise, inconsistency in the raw data exists, which can be identified and eliminated as discussed later. The point is that the final ranking of the alternatives, however it may be generated, must be consistent with the non-conflicting non-outranking relationships contained in N_a.

b) Augmented concordance analysis

If a_k is assumed to be preferred to a_l, or $a_k P a_l$, and if y_j is a concordant attribute of the assumption (i.e. $j \in C_{kl}$), then a question arises as regards the extent to which y_j is concordant with the assumption, or how much support towards $a_k P a_l$ can be gathered from y_j. A further question relates to how to measure the gross support to $a_k P a_l$ if there is more than one attribute concordant with the assumption.

It is necessary that both the multiattribute evaluations and the relative weights be taken into account for such measurement for an unbiased evaluation. It is clear that $|y_{kj} - y_{lj}|$ represents one part of the support to $a_k P a_l$ from y_j in terms of the multiattribute evaluations, and w_j represents another part of the support in terms of the weights. Therefore, a proper sum of $|y_{kj} - y_{lj}|$ for all $j \in C_{kl}$ forms one part of the gross support to $a_k P a_l$ in terms of the multiattribute evaluations, and a proper sum of w_j for all $j \in C_{kl}$ constitutes another part of the gross support to $a_k P a_l$ in terms of the weights.

In fact, in classical concordance and discordance analyses as discussed in [Hwang and Yoon 1981] a concordance index for the assumption $a_k P a_l$ is given by

$$p_{kl} = \sum_{j \in C_{kl}} w_j / S_p \tag{3.85}$$

where S_p is a common scaling factor for all p_{kl} ($k, l \in M$ and $k \neq l$), given by

$$S_p = \sum_{j=1}^{n} w_j \tag{3.86}$$

so that $0 \leq p_{kl} \leq 1$ for any $k, l \in M$ and $k \neq l$. If the weights are normalized as in equations (3.74), it is obvious that $S_p = 1$.

The value of p_{kl} measures the gross importance of the assumption that a_k outranks a_l. Therefore p_{kl} will be called a preference concordance index. Note that $p_{kl}=1$ and $p_{lk}=0$ if a_k happens to be the ideal alternative taking the best values of all the attributes or if a_l happens to be the nadir alternative taking the worst values of all the attributes. Otherwise, $0 < p_{kl} < 1$ for $k, l \in M$; $k \neq l$. Furthermore, $p_{kl} > p_{ij}$ means that the attribute values are more in support of the assumption $a_k P a_l$ than of the assumption $a_i P a_j$.

Similarly, a new evaluation concordance index for $a_k P a_l$, denoted by e_{kl}, may be defined as follows

$$e_{kl} = \sum_{j \in C_{kl}} |r_{kj} - r_{lj}| / S_e \tag{3.87}$$

where r_{ij} is given by equations (3.76) to (3.79). S_e is a common scaling factor for all e_{kl} ($k, l \in M$; $k \neq l$), given by

$$S_e = \sum_{j=1}^{n} \max_{k,l \in M} \{ |r_{kj} - r_{lj}| \} \tag{3.88}$$

so that $0 \le e_{kl} \le 1$ for any $k, l \in M$ and $k \ne l$. It is easy to show that $S_e = n$ if the decision matrix R (see (3.80)) is normalized by equations (3.76) to (3.78).

The value of e_{kl} measures the gross support to the assumption $a_k P a_l$ in terms of the multiattribute evaluations. Note that $e_{kl}=1$ if it happens that a_k is the ideal alternative and a_l is the nadir alternative, and $e_{kl}=0$ if a_k happens to be the nadir or a_l the ideal alternative. Otherwise, $0<e_{kl}<1$ for $k, l \in M$; $k \ne l$. Furthermore, $e_{kl} > e_{ij}$ means that there is greater support for $a_k P a_l$ than that for $a_i P a_j$ in terms of the multiattribute evaluations.

In the same way, a preference concordance index and an evaluation concordance index can be defined for any pair of alternatives. The information obtained by this augmented concordance analysis is summarized by the following two matrices, a preference concordance index matrix PC and an evaluation concordance index matrix EC, which are generally not symmetric

$$PC = \begin{bmatrix} - & p_{12} & \cdots & p_{1m} \\ p_{21} & - & \cdots & p_{2m} \\ p_{31} & p_{32} & \cdots & p_{3m} \\ \cdots & \cdots & \cdots & \cdots \\ p_{m1} & p_{m2} & \cdots & - \end{bmatrix}, \qquad EC = \begin{bmatrix} - & e_{12} & \cdots & e_{1m} \\ e_{21} & - & \cdots & e_{2m} \\ e_{31} & e_{32} & \cdots & e_{3m} \\ \cdots & \cdots & \cdots & \cdots \\ e_{m1} & e_{m2} & \cdots & - \end{bmatrix} \tag{3.89}$$

With the help of the concept of the net dominance relationships, which was introduced by [Van Delft and Nijkamp 1977], a net preference concordance index p_k for an alternative a_k is defined as follows

$$p_k = p(a_k) = \sum_{\substack{l=1 \\ l \ne k}}^{m} (p_{kl} - p_{lk}) \qquad k=1, \cdots, m \tag{3.90}$$

Note that $(p_{kl} - p_{lk})$ stands for the net importance of the assumption $a_k P a_l$. Thus, p_k measures the net importance of the hypothesis that a_k outranks all other competing alternatives. p_k is thus called the net preference concordance index. p_k aggregates the information contained in PC. If the scaling factor defined by equation (3.86) is used, then $-(m-1) \le p_k \le m-1$ for $k=1, \cdots, m$.

Similarly, the following net evaluation concordance index e_k for a_k is defined

$$e_k = e(a_k) = \sum_{\substack{l=1 \\ l \ne k}}^{m} (e_{kl} - e_{lk}) \qquad k=1, \cdots, m \tag{3.91}$$

Note that $(e_{kl} - e_{lk})$ expresses the net support from Table 3.1 towards the assumption $a_k P a_l$. Thus, e_k measures the net support to the hypothesis that a_k

outranks all other competing alternatives in terms of the multiattribute evaluations. e_k aggregates the information contained in EC and represents the net dominance of the alternative a_k in comparison with other alternatives. If the scaling factor defined by equation (3.88) is used, then $-(m-1) \leq e_k \leq m-1$ for $k=1, \cdots, m$.

Note that the process of obtaining p_k and e_k assumes compensation among concordance attributes for each pair of alternatives without any limit. However, this compensation will be modified in the discordance analysis described in the following section so that only compensation within certain limits set by the veto threshold values is permitted.

c) Modified discordance analysis

The evidence against the assumption $a_k P a_l$ must also be measured by means of both the weights and the multiattribute evaluations. However, any index used for such measurement must not be redundant with respect to the preference concordance index p_{kl} and the evaluation concordance index e_{kl} already defined. More important is that certain veto threshold values VT, if provided, have to be taken into account in such an index.

A discordance index is defined as follows,

$$d_{kl} = \begin{cases} \bar{d}_{kl} = \sum\limits_{j \in D_{kl}} \mid z_{kj} - z_{lj} \mid / S_d & \text{if } (a_k, a_l) \notin N_a \\ \bar{d}_{kl} x_{kl}, \quad x_{kl} \geq 0 & \text{if } (a_k, a_l) \in N_a \end{cases} \tag{3.92}$$

where z_{kj} is given by equation (3.81), and x_{kl} is an auxiliary variable which is used to magnify the significance of the evidence objecting to the assumption $a_k P a_l$ up to a point where a_k becomes inferior to a_l. S_d is a common scaling factor for all d_{kl} $(k,l \in M;\ k \neq l)$ given by

$$S_d = \sum_{j=1}^{n} \max_{k,l \in M} \{ \mid z_{kj} - z_{lj} \mid \} \tag{3.93}$$

so that $0 \leq d_{kl} \leq 1$ for any $(a_k, a_l) \notin N_a$. It is easy to show that $S_d=1$ if the decision matrix is normalized by equations (3.76) to (3.78) and the weighted normalized decision matrix Z is obtained by equation (3.81) where w_j is the normalized weight for attribute j given by equation (3.74).

The value of d_{kl} measures the gross objection to the assumption $a_k P a_l$. Note that $\bar{d}_{kl}=1$ if it happens that a_k is the nadir alternative and a_l is the ideal alternative, and that $\bar{d}_{kl}=0$ if a_k happens to be the ideal or a_l the nadir alternative. Otherwise, $0<d_{kl}<1$ for any $(a_k, a_l)\notin N_a$. Furthermore, $d_{kl}>d_{ij}$ means that there is greater objection to the assumption $a_k P a_l$ than that to $a_i P a_j$.

In the same way, a discordance index can be defined for each pair of alternatives. The information obtained by this modified discordance analysis is summarized by the following discordance index matrix CD, which is generally not symmetric

$$CD=\begin{bmatrix} - & d_{12} & \cdots & d_{1m} \\ d_{21} & - & \cdots & d_{2m} \\ d_{31} & d_{32} & \cdots & d_{3m} \\ \cdots & \cdots & \cdots & \cdots \\ d_{m1} & d_{m2} & \cdots & - \end{bmatrix} \tag{3.94}$$

Similar to p_k and e_k, the information contained in CD is aggregated by a net discordance index d_k for an alternative a_k, defined as follows

$$d_k = d(a_k) = \sum_{\substack{l=1 \\ l\neq k}}^{m}(d_{kl} - d_{lk}) \qquad k=1, \cdots, m \tag{3.95}$$

Note that $(d_{kl} - d_{lk})$ denotes the net objection to the assumption $a_k P a_l$. Thus, d_k measures the net objection to the hypothesis that a_k outranks all other competing alternatives. In other words, d_k represents the total inferiority of the alternative d_k in comparison with other alternatives.

d_k and d_l are the functions of x_{kl} for $(a_k, a_l)\in N_a$ where the values of x_{kl} are to be determined to take into account the limited compensation requirement set by equation (3.75). To do so, it is necessary that

$$d_k - d_l \geq \delta \text{ for } any \ (a_k, a_l) \in N_a \tag{3.96}$$

where δ is a small non-negative real number. Equation (3.96) means that the net inferiority of a_l must not be larger than that of a_k if a_k must not outrank a_l.

x_{kl} defined by equation (3.92) is regulated to satisfy equations (3.96). Ideally, x_{kl} should be assigned to 1 or as close to 1 as possible, so that d_{kl} defined by (3.92) for all $k, l\in M$; $k\neq l$ including $(a_k, a_l)\in N_a$ could be calculated in the same way and hence d_{kl} for $(a_k, a_l)\in N_a$ could be as comparable as possible with d_{kl} for $(a_k, a_l)\notin N_a$.

This aspiration and the constraints set by equations (3.96) lead to the construction of the following linear goal programming.

$$\begin{aligned}
\min\ & P_1 \sum_{(a_k,a_l)\in N_a} w_{kl}\zeta_{kl} + P_2 \sum_{(a_k,a_l)\in N_a} w_{kl}(\xi_{kl}^{+} + \xi_{kl}^{-}) \\
s.t.\ & d_k - d_l + \zeta_{kl} \geq \delta \qquad \text{for } all\ (a_k, a_l) \in N_a \\
& x_{kl} + \xi_{kl}^{-} - \xi_{kl}^{+} = 1 \qquad \text{for } all\ (a_k, a_l) \in N_a \\
& \xi_{kl}^{+} \times \xi_{kl}^{-} = 0, \qquad x_{kl}, \zeta_{kl}, \xi_{kl}^{+}, \xi_{kl}^{-} \geq 0
\end{aligned} \tag{3.97}$$

where ζ_{kl}, ξ_{kl}^{+} and ξ_{kl}^{-} are deviation variables and P_1 and P_2 are priority weights with $P_1 \gg P_2$. δ is assigned so that the difference between d_k and d_l is significant. δ may be set so that $1/(10m) \leq \delta \leq 1/m$.

In problem (3.97), w_{kl} is the relative weight of the non-outranking relationship (a_k, a_l) representing the importance of (a_k, a_l) in N_a. w_{kl} is obtained as follows. Let I_{kl} be an index set defined by

$$I_{kl} = \{i \mid y_i(a_l) - y_i(a_k) \geq vt_i \quad \text{for } all\ i \in J\} \tag{3.98}$$

Then, the importance of (a_k, a_l) is given by

$$w_{kl} = \sum_{i \in I_{kl}} w_i \tag{3.99}$$

where w_i is given by equation (3.74). Obviously, $w_{kl} \leq 1$ for any $(a_k, a_l) \in N_a$.

Suppose the optimal solution of problem (3.97) is expressed by $\hat{x}_{kl}$, $\hat{\zeta}_{kl}$, $\hat{\xi}_{kl}^{+}$ and $\hat{\xi}_{kl}^{-}$ for all $(a_k, a_l) \in N_a$. If $\hat{\zeta}_{kl} \leq \delta$ for all $(a_k, a_l) \in N_a$, the raw data is regarded to be consistent and thus $\hat{x}_{kl}$ is used to calculate d_k $(k=1, \cdots, m)$. If any $\hat{\zeta}_{kl} > \delta$, it means that the non-outranking relationships in N_a are conflicting, or the raw data given by Table 3.1, W and VT is inconsistent. In this case, the inconsistency needs to be eliminated. The elimination can be performed by modifying the veto threshold values as follows.

If the number of the non-outranking relationships in N_a is small, the relationships can be displayed to the decision maker who is expected to indicate which ones cause the conflict. Otherwise, all the relationships with $\hat{\zeta}_{kl} > \delta$ are suggested for modification. Suppose (a_k, a_l) is such a non-outranking relationship with its weight w_{kl} being relatively small. Then, the veto threshold values relevant to (a_k, a_l) are modified as follows so that (a_k, a_l) can be dropped from N_a

$$vt_i > y_i(a_l) - y_i(a_k) \quad \text{for } all\ i \in I_{kl} \tag{3.100}$$

The precise values of vt_i need to be assigned by the decision maker by satisfying equation (3.100). In the same way, all such relationships with $\zeta_{kl} > \delta$ can be identified and modified. Let $\overline{N}_a$ be the set of the remaining non-conflicting non-outranking relationships. Then problem (3.97) can be re-solved with N_a being replaced by $\overline{N}_a$ and the new optimal values of x_{kl} for $(a_k, a_l) \in \overline{N}_a$ is then used to calculate d_k for $k=1, \cdots, m$.

3.3.4.3 Preference Matrix and CODASID Algorithm

a) Construction of Preference Matrix

$p(a)$, $e(a)$ and $d(a)$ provide independent information which can be represented by constructing a preference matrix as shown by Table 3.11. $p(a)$, $e(a)$ and $d(a)$ $(a \in A = \{a_k \mid k=1, \cdots, m\})$ are functions obtained by aggregating the original attributes of Table 3.1 through the concordance and discordance analyses, where $p(a)$, $e(a)$ and $d(a)$ represent the net importance, the net dominance and the net inferiority of an alternative a, respectively. Thus, an alternative a_k is more favourable if it has large values of $p(a_k)$ and $e(a_k)$ and a small value of $d(a_k)$.

Table 3.11 *The Preference Matrix*

	$p(a)$	$e(a)$	$d(a)$
a_1	p_1	e_1	d_1
a_2	p_2	e_2	d_2
$\cdots$	$\cdots$	$\cdots$	$\cdots$
a_m	p_m	e_m	d_m

Alternatives are then selected or ranked in the new fixed 3-D space spanned by $p(a)$, $e(a)$ and $d(a)$, which may be referred to as the preference space. As the concordance and discordance analyses process the raw data by comparing the differences in the values of each pair of alternatives on every attribute, it can provide data on the extent to which an alternative either outperforms or underperforms another. As a direct result of the analyses, it is possible that some alternative has better or worse values on all three indices than another alternative. In other words, the former may dominate or be dominated by the latter in the preference space although that is not the case in the original attribute space. Consequently, a partial preference order of this type between some pairs of alternatives is obtained. This partial order always satisfies the limited compensation requirement set by the veto threshold values. The final ranking, however it may be generated, must be consistent with this preference order.

b) Assignment of Trade-off Weights

In the process of ranking of alternatives, the trade-off weight values λ_j for the elements of the preference matrix need to be assigned. If the decision maker decides to gather all pieces of evidence without any bias towards any evidence, which either supports or opposes an assumption that one alternative outranks others, then the following initial relations for λ_1, λ_2 and λ_3 may be given

$$\lambda_1^0 + \lambda_2^0 + \lambda_3^0 = 1 \quad \text{and} \quad \lambda_1^0 + \lambda_2^0 = \lambda_3^0 \tag{3.101}$$

where $\lambda_j = \lambda_j^0$ for $j=1, 2, 3$. Thus

$$\lambda_1^0 + \lambda_2^0 = 0.5, \qquad \lambda_3^0 = 0.5 \tag{3.102}$$

Since $p(a)$ is constructed using pure preference information and $e(a)$ using pure evaluation information, λ_1 and λ_2 actually reflect the relative reliabilities of the multiattribute evaluations and the weights. The assessment of λ_1 and λ_2 mainly depends upon the decision maker's confidence in the two types of information. If both types of information are equally reliable, the same value may then be assigned to λ_1 and λ_2 as follows

$$\lambda_1 = \lambda_2 \quad so \quad \lambda_1^0 = \lambda_2^0 = 0.25 \tag{3.103}$$

The above process of assignment of λ values is of general applicability as the decision maker can always test different sets of preference weights on the original attributes (see equation (3.74)) with the assumption that they are as reliable as the multiattribute evaluations. This strategy is adopted in the second example examined in the later section. However, alternative assumptions about the assignment of the λ values can be accommodated easily. Note that the assignment of λ_1, λ_2 and λ_3 will not change the partial preference order obtained as a direct result of the concordance and discordance analyses.

The assignment given by equations (3.101) and (3.102) will not be justified if the final ranking is inconsistent with the non-conflicting non-outranking relationships given by equation (3.84). This is because the non-outranking relationships are paramount. If such inconsistency occurs, λ's therefore need to be modified as follows

$$\lambda_3^{t+1} = \lambda_3^t + \Delta\lambda_3 \quad \text{and} \quad \lambda_1^{t+1} + \lambda_2^{t+1} = 1 - \lambda_3^{t+1}, \quad t \geq 0 \tag{3.104}$$

with $0 \leq \Delta\lambda_3 \leq 1$. $\Delta\lambda_3$ may be assigned so that

$$\Delta\lambda_3 = \frac{(1 - \lambda_3^0)}{N_\lambda} \tag{3.105}$$

where N_λ is an integer with $10 \le N_\lambda \le 100$. When $\lambda_1^{t+1} + \lambda_2^{t+1} \le \Delta\lambda_3$, however, let $\lambda_3^0 = \lambda_3^t$ and then re-calculate $\Delta\lambda_3$ and λ_3^{t+1} as well as λ_1^{t+1} and λ_2^{t+1}.

As λ_3 increases, more importance is shifted to the net discordance index $d(a)$ which provides a preference order consistent with the non-conflicting non-outranking relationships, as discussed in the last section. Therefore, increasing λ_3 actually means that the limited compensation given by equations (3.75) and (3.84) is forced to take place. It is easy to show that this limited compensation can always be satisfied as λ_3 becomes sufficiently large ($\lambda_3 \le 1$).

c) Computational Steps

The computational procedure of CODASID can be summarized as follows.

Step 1: Model a design selection problem in terms of a decision matrix (Table 3.1), weights and possibly veto threshold values for attributes.

Step 2: Generate the normalized decision matrix (3.80) using the technique described by equations (3.76) to (3.78), obtain the weighted normalized decision matrix (3.81), and generate the non-outranking relationships (3.84).

Step 3: For each pair of alternatives (a_k, a_l) $(k, l=1, \cdots, m; k \ne l)$, construct the concordance set (3.82), and calculate the preference concordance index (3.85) and the evaluation concordance index (3.87). Formulate the preference concordance index matrix and the evaluation concordance index matrix (3.89). Then, calculate the net preference concordance index p_k (3.90) and the net evaluation concordance index e_k (3.91) for $k=1, \cdots, m$.

Step 4: For each pair of alternatives, construct the discordance set (3.82) and calculate the discordance index (3.92). Formulate the discordance index matrix CD (3.94) and calculate the net discordance index d_k (3.95) for $k=1, \cdots, m$. Then, find the optimal setting of d_k for $k=1, \cdots, m$ using the linear goal programming formulation (3.97) by taking into account the veto threshold values. In this step, some veto threshold values may need to be modified if inconsistency occurs in the raw data, as discussed in Section 3.3.4.1-c.

Step 5: Summarize the aggregated information by constructing the preference matrix and identify those possible preference relationships with one alternative dominating or being dominated by another in the preference space. Let $t=0$ and assign the initial trade-off weights λ's by equations (3.101) to (3.103).

Step 6: Based on Table 3.11, rank the alternative designs using the TOPSIS method.

Step 7: If the ranking of the alternatives is inconsistent with the non-conflicting non-outranking relationships (3.84), let $t=t+1$, regulate the trade-off weights using equations (3.104) and (3.105) and then go to step 6 while the normalized preference matrix need not be re-calculated. Otherwise, go to step 8.

Step 8: An alternative with the largest value of the index is the most preferred alternative, and the others can be ranked accordingly.

3.3.4.4 Applications

a) Fighter Aircraft Selection

Let's take for example the fighter aircraft selection problem, as shown by Table 3.10, to demonstrate the CODASID method. In addition to Table 3.10 and equation (3.66), veto threshold values for the six attributes are given by

$$VT = [0.65 \quad 1.5 \quad 1.0 \quad 2.5 \quad 5.0 \quad 5.0]^T \tag{3.106}$$

which indicates that aeroplane i (a_i) must never be preferred to aeroplane j (a_j) if a_j is faster than a_i by 0.65 Mach Number, or if a_j can fly 1500 nautical miles longer, or if a_j can carry 1000 more pounds of ammunition, or if a_j is 2.5 million pounds cheaper, or if a_j is far more reliable (5 units of difference) or if a_j has far better manoeuvrability (5 units of difference).

Comparing VT with the data in Table 3.10, one can find that only one non-outranking relationship among the feasible alternatives can be detected, that is, aeroplane 3 must never be preferred to aeroplane 2, or (a_3, a_2). This is because a_2 is faster than a_3 by 0.7 Mach Number, beyond the limit of 0.65 Mach Number as stated in (3.106). Thus

$$N_a = \{(a_3, a_2)\} \tag{3.107}$$

To explain how the new concordance and discordance analyses gather evidence from the raw data, let's take for example the assumption that the aeroplane a_1 is better than a_2. In this case, the evidence supporting the assumption includes the facts that a_1 can carry 200 more pounds of ammunition than a_2, is a million pounds cheaper, more reliable (2 units of difference) and much more manoeuvrable (4 units of difference). Besides, the gross importance of the assumption is 0.607. The evidence objecting to the assumption includes that a_2 is faster than a_1 by 0.5 Mach Number and a_2 can fly 1200 nautical miles longer. Both types of evidence will then be used to calculate the three new indices, as discussed later. Similar evidence can also be gathered for other assumptions. However, equation (3.107) indicates that a_3 must never be more desirable than

a_2 notwithstanding other evidence that may be gathered.

Following the computational steps listed in the last section, the computation for this example can then be summarized as follows.

Step 2: The normalized decision matrix R and weighted normalized decision matrix Z using the data from equations (3.65) and (3.67)

$$R = \begin{bmatrix} 0.3846 & 0.2500 & 0.5000 & 0.5000 & 0.5000 & 0.9000 \\ 0.7692 & 0.8500 & 0.3000 & 0.1667 & 0.3000 & 0.5000 \\ 0.2308 & 0.5000 & 0.6000 & 0.8333 & 0.7000 & 0.7000 \\ 0.5385 & 0.4000 & 0.5000 & 0.6667 & 0.5000 & 0.5000 \\ 1.0000 & 1.0000 & 1.0000 & 1.0000 & 1.0000 & 1.0000 \\ 0.0000 & 0.0000 & 0.0000 & 0.0000 & 0.0000 & 0.0000 \end{bmatrix} \qquad (3.108)$$

$$Z = \begin{bmatrix} 0.1236 & 0.0179 & 0.0357 & 0.0536 & 0.0536 & 0.2893 \\ 0.2473 & 0.0607 & 0.0214 & 0.0179 & 0.0321 & 0.1607 \\ 0.0742 & 0.0357 & 0.0429 & 0.0893 & 0.0750 & 0.2250 \\ 0.1731 & 0.0286 & 0.0357 & 0.0714 & 0.0536 & 0.1607 \\ 0.3214 & 0.0714 & 0.0714 & 0.1071 & 0.1071 & 0.3214 \\ 0.0000 & 0.0000 & 0.0000 & 0.0000 & 0.0000 & 0.0000 \end{bmatrix} \qquad (3.109)$$

Step 3: The preference concordance index matrix PC, the evaluation concordance index matrix EC

$$PC = \begin{bmatrix} - & 0.6071 & 0.6429 & 0.5000 & 0.0000 & 1.0000 \\ 0.3929 & - & 0.3929 & 0.7143 & 0.0000 & 1.0000 \\ 0.3571 & 0.6071 & - & 0.6786 & 0.0000 & 1.0000 \\ 0.6786 & 0.6071 & 0.3214 & - & 0.0000 & 1.0000 \\ 1.0000 & 1.0000 & 1.0000 & 1.0000 & - & 1.0000 \\ 0.0000 & 0.0000 & 0.0000 & 0.0000 & 0.0000 & - \end{bmatrix} \qquad (3.110)$$

$$EC = \begin{bmatrix} - & 0.1889 & 0.0590 & 0.0667 & 0.0000 & 0.5058 \\ 0.1641 & - & 0.1481 & 0.1135 & 0.0000 & 0.4810 \\ 0.1472 & 0.2611 & - & 0.1278 & 0.0000 & 0.5940 \\ 0.0784 & 0.1500 & 0.0513 & - & 0.0000 & 0.5175 \\ 0.4942 & 0.5190 & 0.4060 & 0.4825 & - & 1.0000 \\ 0.0000 & 0.0000 & 0.0000 & 0.0000 & 0.0000 & - \end{bmatrix} \qquad (3.111)$$

Step 4: Since there is only one relation in N_a, that is (a_3, a_2), we can obtain the following discordance index matrix CD.

$$CD = \begin{bmatrix} - & 0.1665 & 0.0821 & 0.0780 & 0.4264 & 0.0000 \\ 0.2000 & - & 0.2000 & 0.0893 & 0.4599 & 0.0000 \\ 0.1137 & 0.1981x_{32} & - & 0.0989 & 0.4580 & 0.0000 \\ 0.1286 & 0.1063 & 0.1179 & - & 0.4769 & 0.0000 \\ 0.0000 & 0.0000 & 0.0000 & 0.0000 & - & 0.0000 \\ 0.5736 & 0.5401 & 0.5420 & 0.5231 & 1.0000 & - \end{bmatrix} \quad (3.112)$$

From equations (3.95) and (3.112), we have $d_2 = 0.1363 - 0.1981x_{32}$ and $d_3 = 0.1981x_{32} - 0.2714$. Note that from equations (3.67), (3.98), (3.99) and (3.107), we get $w_{32}=w_1=0.3215$. Let $\delta = 1/(10\times 6) = 0.0167$. Problem (3.97) is thus formulated by

$$\begin{aligned} \min \quad & P_1[0.3215\zeta_{32}] + P_2[0.3215(\xi_{32}^+ + \xi_{32}^-)] \\ s.t. \quad & 0.3962x_{32} - 0.4077 + \zeta_{32} \geq 0.0167 \\ & x_{32} + \xi_{32}^- - \xi_{32}^+ = 1 \\ & \xi_{32}^+ \times \xi_{32}^- = 0, \qquad x_{32}, \zeta_{32}, \xi_{32}^+, \xi_{32}^- \geq 0 \end{aligned} \quad (3.113)$$

It is easy to show that the optimal solution of the above problem is $\hat{x}_{32}=1.0712$, $\hat{\xi}_{32}^+=0.0712$, $\hat{\zeta}_{32}=\hat{\xi}_{32}^-=0$. We can then obtain $\hat{d}_2=-0.0759$ and $\hat{d}_3=-0.0592$.

Step 5: The preference matrix PM can then be constructed from equations (3.110) to (3.112) as follows using equations (3.90), (3.91) and (3.95)

$$PM = \begin{bmatrix} 0.3214 & -0.0637 & -0.2629 \\ -0.3214 & -0.2124 & -0.0759 \\ 0.2857 & 0.4658 & -0.0592 \\ -0.2857 & 0.0068 & 0.0404 \\ 5.0000 & 2.9017 & -2.8212 \\ -5.0000 & -3.0983 & 3.1788 \end{bmatrix} \quad (3.114)$$

Note that a^* still dominates and a^- is dominated by all other alternatives in the preference space.

Step 6: The normalized preference matrix $\overline{PM}$ and the weighted normalized preference matrix $\tilde{PM}$ are given by

$$\overline{PM} = \begin{bmatrix} 0.0453 & -0.0149 & -0.0617 \\ -0.0453 & -0.0497 & -0.0178 \\ 0.0403 & 0.1089 & -0.0139 \\ -0.0403 & 0.0016 & 0.0095 \\ 0.7045 & 0.6786 & -0.6623 \\ -0.7045 & -0.7246 & 0.7463 \end{bmatrix} \tag{3.115}$$

$$\tilde{PM} = \begin{bmatrix} 0.0453\lambda_1 & -0.0149\lambda_2 & -0.0617\lambda_3 \\ -0.0453\lambda_1 & -0.0497\lambda_2 & -0.0178\lambda_3 \\ 0.0403\lambda_1 & 0.1089\lambda_2 & -0.0139\lambda_3 \\ -0.0403\lambda_1 & 0.0016\lambda_2 & 0.0095\lambda_3 \\ 0.7045\lambda_1 & 0.6786\lambda_2 & -0.6623\lambda_3 \\ -0.7045\lambda_1 & -0.7246\lambda_2 & 0.7463\lambda_3 \end{bmatrix} \tag{3.116}$$

The best preferred and least preferred design are given by a^* and a^-. Thus

$$[\tilde{p}(a^*) \;\; \tilde{e}(a^*) \;\; \tilde{d}(a^*)]^T = [0.7045\lambda_1 \quad 0.6786\lambda_2 \quad -0.6623\lambda_3]^T \tag{3.117}$$

$$[\tilde{p}(a^-) \;\; \tilde{e}(a^-) \;\; \tilde{d}(a^-)]^T = [-0.7045\lambda_1 \quad -0.7246\lambda_2 \quad 0.7463\lambda_3]^T \tag{3.118}$$

Step 7: λ's are regulated with N_λ=10 following the procedure as described in Section 3.3.4.3-b, so that the limited compensation set by equation (3.106) is just satisfied strictly. This regulation process results in $\lambda = \lambda^6 = [\lambda_1^6 \;\; \lambda_2^6 \;\; \lambda_3^6]^T = [0.1 \quad 0.1 \quad 0.8]^T$ where the relative closeness indices are given by

$$[u(a_1) \;\; u(a_2) \;\; u(a_3) \;\; u(a_4) \;\; u(a^*) \;\; u(a^-)]^T$$

$$= [0.5719 \quad 0.5404 \quad 0.5403 \quad 0.5222 \quad 1.0000 \quad 0.0000]^T \tag{3.119}$$

with $u(a_2) > u(a_3)$. Although the difference between $u(a_2)$ and $u(a_3)$ is marginal, it is sufficient to satisfy the requirement represented by equation (3.107).

Step 8: Since $u(a^*) > u(a_1) > u(a_2) > u(a_3) > u(a_4) > u(a^-)$, the ranking of the four alternatives is

$$a^* \succ a_1 \succ a_2 \succ a_3 \succ a_4 \succ a^- \tag{3.120}$$

It is easy to prove that the matrix *PM* given by equation (3.114) has a rank of 3, which indicates that each of the three new indices $p(a)$, $e(a)$ and $d(a)$ process the raw information in an independent manner. Note that from equation (3.114) the following partial preference order is generated for the feasible alternatives

$$\{a_1 \succ a_2; \quad a_3 \succ a_4\} \tag{3.121}$$

as a direct result of the concordance and discordance analyses. This partial order is not changed in the regulation of λ's.

It is of interest to examine the same problem using other existing simple methods. The simple weighting method gives the following results when the weighted normalized matrix given by equation (3.109) is used

$$\begin{aligned}&[u(a_1) \;\; u(a_2) \;\; u(a_3) \;\; u(a_4) \;\; u(a^*) \;\; u(a^-)]^T \\ &\quad = [0.5736 \quad 0.5401 \quad 0.542 \quad 0.5231 \quad 1.0000 \quad 0.0000]^T\end{aligned} \tag{3.122}$$

$$a^* \succ a_1 \succ a_3 \succ a_2 \succ a_4 \succ a^- \tag{3.123}$$

This problem was dealt with using the TOPSIS method in the last Section 3.3.3. As shown by equations (3.72) and (3.73), TOPSIS generates the following results

$$\begin{aligned}&[u(a_1) \;\; u(a_2) \;\; u(a_3) \;\; u(a_4) \;\; u(a^*) \;\; u(a^-)]^T \\ &\quad = [0.7243 \quad 0.5110 \quad 0.6272 \quad 0.5051 \quad 1.0000 \quad 0.0000]^T\end{aligned} \tag{3.124}$$

$$a^* \succ a_1 \succ a_3 \succ a_2 \succ a_4 \succ a^- \tag{3.125}$$

Both the methods provide the same ranking as shown by equations (3.123) and (3.125), which is obviously wrong as a_3 must not be preferred to a_2 in the context of veto threshold values. This is because these two methods both assume unlimited compensation. It would be reasonable to rank a_3 over a_2 without consideration of the veto values (see equation (3.106)) as the former is more attractive in terms of the last four attributes in Table 3.10. In fact, the new method would generate the same ranking as those given by equations (3.123) and (3.125) if equation (3.106) is not considered, that is

$$\begin{aligned}&[u(a_1) \;\; u(a_2) \;\; u(a_3) \;\; u(a_4) \;\; u(a^*) \;\; u(a^-)]^T \\ &\quad = [0.5553 \quad 0.5182 \quad 0.5483 \quad 0.5135 \quad 1.0000 \quad 0.0000]^T\end{aligned} \tag{3.126}$$

$$a^* \succ a_1 \succ a_3 \succ a_2 \succ a_4 \succ a^- \tag{3.127}$$

where λ's are assigned by equations (3.101) to (3.103).

The above comparative study shows the inability of the simple weighting method and the TOPSIS method to deal with such selection problems that require limited compensation among attributes. In such circumstances, the advantage of the new method is clear. In other cases, the new method is capable of generating results comparable to those obtained using existing methods with unlimited compensation.

b) A Cargo Ship Selection Problem

This is a selection problem with 6 competing cargo ship designs [Sen 1988], which are to be compared on the basis of 23 attributes shown in Table 3.12. These attributes were either obtained from published sources (e.g. route independent characteristics like speed, capital cost of ship etc) or those obtained on the basis of operational simulation (e.g. days between calls). The decision matrix is listed in Table 3.12. The first 7 attributes are for maximization and the rest for minimization. The weights in Table 3.13 were chosen by the decision maker to reflect a range of value systems. The decision maker has recognized that any attribute can be offset by others without any limit. Thus, no veto threshold value for any attribute is provided in this example.

To see how the computation procedure works, only part of the procedure for Value System 1 is illustrated below.

Step 5: The preference matrix PM is obtained as follows

$$PM = \begin{bmatrix} -0.0534 & 0.0028 & 0.2364 \\ -0.6868 & -0.5410 & 0.5018 \\ -1.9395 & -1.0071 & 1.1784 \\ 2.2456 & 1.4809 & -1.6555 \\ -0.1459 & -0.0145 & -0.0740 \\ 0.5801 & 0.0790 & -0.1870 \end{bmatrix} \tag{3.128}$$

Step 6: The relative closeness indices are generated by

$$[u(a_1)\ \ u(a_2)\ \ u(a_3)\ \ u(a_4)\ \ u(a_5)\ \ u(a_6)]^T$$
$$= [0.3657\ \ 0.2420\ \ 0.0000\ \ 1.0000\ \ 0.4327\ \ 0.4948]^T \tag{3.129}$$

Step 8: The final ranking of the six cargo ship designs for Value System 1 is given by

Table 3.12 *A Cargo Ship Selection Problem*

No	Attributes	Ship 1	Ship 2	Ship 3	Ship 4	Ship 5	Ship 6
1	*Load Factor	0.70	0.65	0.68	0.70	0.66	0.60
2	*Bale Cap.(m^3)	21070	28000	31587	26718	28874	33117
3	*TEU Capacity	558	704	744	716	1160	1200
4	*Deadweight (t)	17200	23200	22351	20621	22233	25500
5	*Speed (kn)	14.3	14.8	17.7	15.0	18.2	17.6
6	*Secondhand price (pounds M)	3.33	3.0	6.26	3.66	6.25	6.25
7	*Effective load factor (%)	95	93	92	95	90	88
8	Service Route 1 (day between calls)	25.23	26.81	67.40	24.18	34.00	35.17
9	Service Route 2 (day between calls)	8.35	9.14	23.34	8.01	11.93	12.30
10	Days in each port	2.0	2.7	2.2	1.9	2.4	2.5
11	Cap. Inv. per ship (pounds M) (p)	16.66	20.00	31.33	16.66	25.00	25.00
12	Present Value of P(%) (given loan terms)	88	88	88	88	88	88
13	Annual Manning Costs (pounds M)	0.7	0.75	0.80	0.75	0.85	0.85
14	Sea Fuel Cons (tons per day)	18	23	37	20.2	38	38.5
15	Port Fuel Cons (tons per day)	2.1	2.3	2.0	1.9	2.0	2.0
16	Annual M&R Costs (pounds M)	0.25	0.255	0.245	0.22	0.25	0.24
17	Off-hire (days per year)	15	16	18	18	19	17
18	RFR1 (Route 1 per m^3)	32.18	31.96	32.19	24.73	31.37	30.88
19	RFR2 (Route 1 per TEU)	1215	1271	1367	923	781	852
20	RFR3 (Route 1 per dwt)	39.42	38.57	45.49	32.04	40.74	40.10
21	RFR4 (Route 2 per m^3)	10.80	10.95	11.11	8.33	10.93	10.72
22	RFR5 (Route 2 per TEU)	408	436	472	311	272	296
23	RFR6 (Route 2 per dwt)	13.24	13.22	15.70	10.79	14.20	13.92

TEU: Twenty foot equivalent unit; or a standard basic container

RFR: Required freight rate; or the cost of transportation

$$a_4 \succ a_6 \succ a_5 \succ a_1 \succ a_2 \succ a_3 \tag{3.130}$$

It may be of interest to note from (3.128) that ship 4 dominates and ship 3 is dominated by all the other ships in the preference space for Value System 1 although no ship dominates or is dominated by the others in the attribute space defined by Table 3.12. Thus, ship 4 and ship 3 are undoubtedly ranked to be the best and the worst compromise choices for Value System 1, respectively. Furthermore, as a direct result of the concordance and discordance analyses, the following partial preference order for Value System 1 is obtained from (3.128)

$$\begin{aligned}\{a_4 \succ a_1, a_2, a_3, a_5, a_6;\ a_6 \succ a_1, a_2, a_3, a_5;\\ a_5 \succ a_2, a_3;\ a_1 \succ a_2, a_3;\ a_2 \succ a_3\}\end{aligned} \tag{3.131}$$

Thus, only the relation between a_1 and a_5 remains unassigned after the concordance and discordance analyses have been conducted for Value System 1. This unknown relation is then made clear by the alternative ranking process, which shows $a_5 \succ a_1$ with $\lambda_1 = \lambda_1 = 0.25$ and $\lambda_3 = 0.5$.

Table 3.13 *Weights* for *Cargo Ship Selection*

No. of Attributes	Value System 1	Value System 2	Value System 3	Value System 4
1	6	8	3	7
2	7	9	10	4
3	7	9	1	4
4	7	9	10	4
5	4	6	2	3
6	4	6	2	7
7	5	8	3	8
8	3	7	2	9
9	3	7	2	9
10	2	6	5	4
11	8	10	10	9
12	8	10	10	9
13	6	3	5	2
14	5	8	2	8
15	5	8	2	8
16	4	5	5	2
17	4	1	5	2
18	10	5	1	5
19	10	5	1	5
20	10	5	1	5
21	7.5	5	1	10
22	7.5	5	1	10
23	7.5	5	1	10

Table 3.14 *Relative Closeness Indices to The Ideal Point*

ship designs	relative closeness indices			
	Value System 1	Value System 2	Value System 3	Value System 4
ship 1	0.3657	0.3472	0.3863	0.4886
ship 2	0.2420	0.1601	0.2758	0.2655
ship 3	0.0000	0.0000	0.0638	0.0000
ship 4	1.0000	1.0000	1.0000	1.0000
ship 5	0.4327	0.4172	0.1791	0.4282
ship 6	0.4948	0.4342	0.4827	0.4060

Table 3.15 *Rankings of The Cargo Ship Designs*

ship designs	rankings			
	Value System 1	Value System 2	Value System 3	Value System 4
ship 1	4	4	3	2
ship 2	5	5	4	5
ship 3	6	6	6	6
ship 4	1	1	1	1
ship 5	3	3	5	3
ship 6	2	2	2	4

The results for all the four value systems are summarized in Table 3.14 and Table 3.15. From these tables, it is clear that regardless of the changes of the value systems ship 4 is always considered to be the most preferred design and ship 3 is the worst one. This conclusion is consistent with the suggestion proposed in reference [Sen 1992] using other techniques.

c) Design Selection of Semi-Submersibles

Finally, let's examine the selection problem of the 13 efficient designs for the semi-submersible, as shown by Table 3.3 and by equation (3.37). This example was examined earlier by using the TOPSIS method. Using the CODASID method, however, we can obtain the following closeness indices of the designs

$$[u(a_1)\ \ u(a_2)\ \ u(a_3)\ \ u(a_4)\ \ u(a_5)\ \ u(a_6)$$
$$u(a_7)\ \ u(a_8)\ \ u(a_9)\ \ u(a_{10})\ \ u(a_{11})\ \ u(a_{12})\ \ u(a_{13})]$$
$$= [0.8198\ \ 0.3945\ \ 0.3771\ \ 0.5569\ \ 0.6143\ \ 0.2662\ \ 0.8255$$

0.8127 0.7744 0.9526 0.9224 0.9341 0.8448]

which results in the following ranking of these designs

$$a_{10} \succ a_{12} \succ a_{11} \succ a_{13} \succ a_7 \succ a_1 \succ a_8 \succ a_9 \succ a_5 \succ a_4 \succ a_2 \succ a_3 \succ a_6$$

a_{10} is ranked as the best design by the CODASID method as it is the closest to the ideal design. a_6 is regarded as the worst design as its performance is rather poor over the first five attributes even though it is the cheapest design.

3.3.5 Comments

It is clear on the basis of the above that alternative methods of communicating the design data, the preference structure of the decision maker and the decision rule governing the processing of the data can and does lead to alternative ranking of the available alternatives. This is only to be expected as alternative formulations and solution strategies represent alternative points of view. The decision maker has the task of matching methods with the problem in hand and it is this creative nature of the task that makes multiple attribute decision making in design a meaningful and important activity.

3.4 A Hierarchical Evaluation Process

3.4.1 Design Decision Problems with Subjective Factors

In the realm of engineering design, techno-economic evaluations are often used to assess the suitability of alternative technical solutions. For example, the choice of a particular ship or fleet of ships for a given transportation function is often based on some appropriate economic criterion. Such assessments usually deal only with objective information about vessel performance, based on some idealised or simplified route characteristics. However, it is recognised that the idealised study only represents some notional averaged conditions of operation and, in fact, the numerical assessments are only meant to provide the data on which the designer's judgements can be based. This is to take account of variations in operating conditions and for the consideration of subjective factors that cannot be conveniently modelled in the context of techno-economic evaluations.

It would appear, therefore, that if subjective and objective assessments could be considered together within a rational decision making process, decisions on design could be made in the confidence that the relative influence of all factors could be satisfactorily represented and evaluated. This section discusses one technique to such problems based on an evidential reasoning approach [Yang and

Singh 1994a][Yang and Sen 1994b].

a) The candidate designs

The design study that is addressed in this section deals with the problem of choosing one of a number of standard designs for a certain range of transportation tasks. To keep the discussion managable, the assessment is limited to three designs. The principal characteristics of these three designs are outlined in Table 3.16, as obtained from [Buxton 1989].

Table 3.16 *Candidate Designs*

Type	Ship 1	Ship 2	Ship 3
Initial price	16.66	25.00	16.66
Speed	15.00	17.60	14.30
Deadweight (tonnes)	20621	25500	17200
Cubic capacity, bale (m^3)	26718	33117	21070
Container number (TEU)	716	1200	558
Engine power (kw)	4890	10640	3970
Main engine fuel (tonne/day)	19.0	37.0	18.0
Auxiliary fuel, sea/port (t/day)	1.2/1.9	1.5/2.0	0/2.1
Cargo gear (cranes × tonnes)	4 ×25	4 × 23.5	5 × 10

In brief, Ship 1 is the KOTA SINGA from Mitsubishi Heavy Industries Ltd. It can carry general cargo, dry bulk cargo, containers and long-sized cargoes like steel pipes. It has slow speed diesels and four cranes.

Ship 2 is similar to a Container / Multi purpose Carrier from Thyssen Nordseewerke, being a 25,500 dwt, 1200 TEU ship and is container oriented to the extent that tween-deck hatch covers are not offered as standard. It has a slow speed diesel and a faster speed than Ship 1.

Ship 3 is taken to be similar to a Freedom Mark II, a multi-purpose general cargo, container and self-trimming dry cargo carrier of 17000 dwt from Ishikawajima - Harima Heavy Industries Co. Ltd. The vessel is equipped with retractable tween decks and horizontal slewing cargo gear. It is equipped with medium speed engines.

Although the physical characteristics of the vessels are as described above, the assessments that follow are on behalf of a hypothetical owner over a range of operating scenarios as described below. The evaluations are meant to show how subjective and objective factors can be considered together and must not be viewed as anything other than demonstrative.

b) The operating scenarios

The ships in question are being considered for operation over two different routes and the range of operating conditions encountered are idealised into three operating scenarios. In reality, a larger number of operating scenarios would probably need to be considered but the designer or operator can only respond meaningfully to a certain maximum level of complexity and this has to be borne in mind. In practice alternative sets of weightings for these scenarios can be used to create other operating profiles.

The scenarios considered are as follows :

Scenario 1 : A containerised route favouring higher speed vessels.

Scenario 2 : Roughly split between break bulk and containerised cargo.

Scenario 3 : Mixed cargo on a relatively shorter route compared to Scenarios 1 and 2.

For purposes of practical assessment it is assumed that the annual transported quantity is the same for all the vessels. Larger faster ships can carry more cargo each time and hence tend to offer a less frequent service although each unit of cargo can be expedited.

The transportation task is such that it favours container friendly vessels of modest size and speed. Competitors in the route in question are assumed to be operating largely multi-purpose ships with speed range of 14 to 15 knots.

c) The evaluation factors

The evaluation factors on the basis of which assessments are to be made can broadly be seen as either objective or subjective. Subjective judgements can only be taken into account using a process that explicitly takes account of the imprecise nature of such judgements. How this may be done using an evidential reasoning approach is addressed in Section 3.4.2 but the factors themselves are outlined below.

Objective factors :

- Ship price
- Ship speed

Subjective factors :

- Cargo factors
- Economic factors
- Service factors

These compound subjective factors (referred to as attributes) need to be further decomposed into factors to which the designer or operator can relate. The sub-factors are as follows :

Cargo factors :

- Bale utilization
- TEU utilization
- Dwt utilization

Economic factors :

- Maintenance & repair
- Relative fuel consumption
- Relative freight cost
- Resale potential

Service factors :

- Frequency of service
- Cargo expedition

The cargo factors deal with the representation of utilization of cargo space. The economic factors are concerned with some of the factors that normally go into the techno-economic assessments of technical alternatives. The service factors deal with aspects of the service that customers are often interested in but that are usually only implicitly addressed in formal assessments.

It will be noted that factors for evaluation need not be totally independent of each other. It is, however, necessary for assessments to be consistent in that where some evaluation factors are closely related to one another the subjective judgements over those factors need to be non-contradictory. If the judgements are real, this should not pose any problem.

The hierarchical arrangement of factors could be extended to whatever level of decomposition that may be deemed to be necessary for a given problem. However, for this problem the above level of decomposition is considered to be adequate. The hierarchy of subjective factors is therefore as shown in Figure

3.15.

The next section describes how subjective judgements can be represented and combined with objective data within a hierarchical framework to lead to a ranking of the candidate designs.

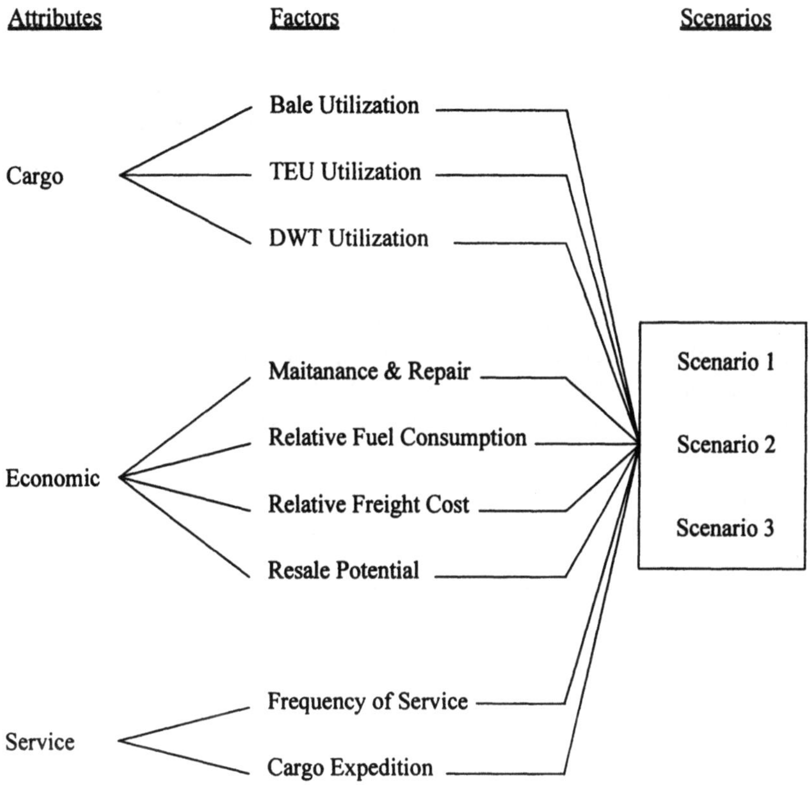

Figure 3.15. *Hierarchy of Factors*

3.4.2 A Hierarchical Evaluation Process

a) Subjective judgements and evidential reasoning

As indicated in Figure 3.15, the state of a compound factor (an attribute) may be determined by factors at a lower level. If an attribute is only associated with one factor whose state is judged to be absolutely "*Good*", for example, then the state of the attribute must also be "*Good*". Here "*Good*" stands for an evaluation grade, indicating a distinct state of a factor. Other evaluation grades such as "*Poor*", "*Average*", "*Excellent*" may be used as well. Generally, an attribute may have several factors associated with it. If the states of the factors are all evaluated to be absolutely "*Good*", then the state of the associated attribute should certainly be "*Good*".

However, such certain and consistent evaluations are rarely available in practice. Indeed, judgements are often uncertain and multiple evaluation grades may be used in a single judgement when several factors need to be taken into account simultaneously. It could be judged, for example, that for scenario 1 the *Frequncey of Service* of *Ship* 1 is "*Very Good*" and "*Good*" with the degrees of confidence (belief) of 0.4 and 0.6 respectively and its *Cargo Expedition* is "*Excellent*", "*Very Good*" and "*Good*" to the extents of 0.5, 0.2 and 0.2 respectively. In the above judgements, the real numbers such as 0.2, 0.4, 0.5 and 0.6 represent uncertainty in the evaluations which is caused by the complexity of the factors as well as the inability of the designer to give precise evaluations. It may be noted that the sum of the confidence degrees in a judgement for a factor may not necessarily be equal to one. This indicates incomplete uncertainty where the remaing uncertainty is unassigned due to lack of evidence.

Problems may then arise as to how the above uncertain evaluations of factors (such as *Frequncey of Service* and *Cargo Expedition*) may be synthesized in a rational way so as to attain an (often uncertain) evaluation for the associated attribute (such as *Service*). The problem may be generalized to one of addressing how each attribute could be measured through the evaluations of these factors by taking consideration of all the three scenarios. The hierarchical evaluation process provides an approach for dealing with such synthesis problems.

The evaluation process is based on the Dempster-Shafer (simply D-S) theory which is well suited for handling incomplete uncertainty. This theory can model the narrowing of the hypothesis set with the accumulation of evidence. In other words, the chance that a hypothesis is true will be improved if more pieces of evidence support the hypothesis. In Figure 3.15, whether the state of a factor for all the three scenarios is "*Good*" would be regarded as a hypothesis. The given judgement about the state of a factor for one scenario may be viewed as a single piece of evidence. The hierarchical evaluation process provides a systematic way of synthesizing uncertain evaluations of a factor for each scenario to produce an evaluation of the factor for all the three scenarios.

To apply the D-S theory, the mutual exclusiveness and exhaustiveness of all hypotheses have to be satisfied. It is therefore necessary that all the adopted evaluation grades be defined as distinct grades. In other words, if one of the grades is absolutely confirmed, all the other grades must not be confirmed at all; if more than one grade is confirmed simultaneously, the total degree of confidence must be one or smaller than one. In addition to this requirement, the grades must cover all possible grades the designer may need to use to judge the states of factors.

b) Basic evidential reasoning model and algorithm

An evaluation analysis model is used as a framework in which to conduct multiple factor analysis and reasoning with uncertain decision knowledge. A basic evaluation analysis model may be constructed as shown in Figure 3.16, in which only a single level of factors are involved.

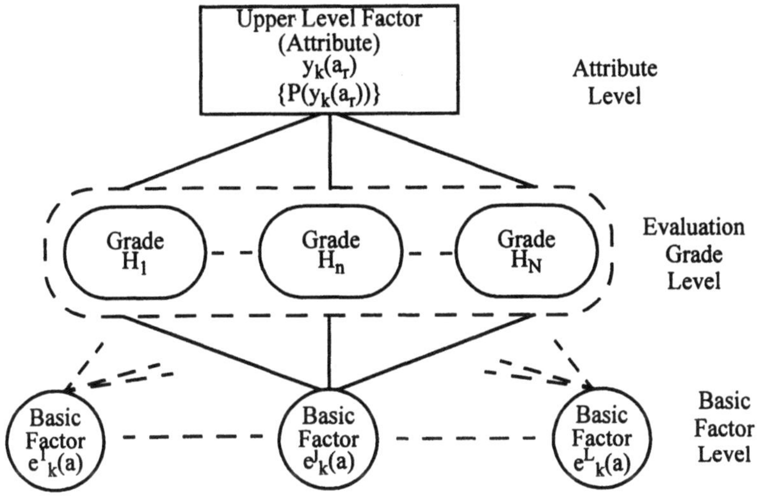

Figure 3.16. *A Basic Evaluation Model*

In Figure 3.16, e_k^j denotes a factor such as *Bale Utilization*, which can be directly evaluated for a given ship at a given *Scenario*. The set of factors for evaluation of an attribute, denoted by y_k, is defined by

$$E_k = \{e_k^1 \cdots e_k^j \cdots e_k^{L_k}\} \tag{3.132}$$

$m_{kj}^n = m(H_n/e_k^j(a))$ expresses a basic probability assignment to which a factor e_k^j supports a hypothesis that the state of an attribute y_k at a ship (an option) a is confirmed to H_n. m_{kj}^n can be generated from the given confidence degree and the normalized relative weight of e_k^j. $m_k^n = m(H_n/E_k(a))$ represents an overall probability assignment to which the state of y_k at a is confirmed to H_n by the whole factor set E_k. m_k^n is obtained by combining all m_{kj}^n (j=1, $\cdots$, L_k;

$n=1, \cdots, N$).

Suppose $\beta_{kj}^{n}(a_r)$ denotes a confidence degree associated with the state of a factor e_k^j at a ship a_r being evaluated to H_n. Then, an uncertain subjective judgement for evaluation of the state of $e_k^j(a_r)$) may be expressed by the following expectation

$$S(e_k^j(a_r)) = \{(\beta_{kj}^{n}(a_r), H_n),\ \ n=1, \cdots, N;\ \text{and} \sum_{n=1}^{N} \beta_{kj}^{n}(a_r) \le 1\} \quad (3.133)$$

which indicates that the state of e_k^j at a design a_r is evaluated to H_n with a confidence degree of $\beta_{kj}^{n}(a_r)$ for $n=1, \cdots, N$. In the above formula, we assume that the state of a basic factor e_k^j at a_r may be evaluated to any individual evaluation grade defined in H. More general uncertain subjective judgements can be handled as well, as discussed in [Yang and Singh 1994a][Yang and Sen 1994b].

$S(e_k^j(a_r))$ may then be quantified using its preference degree, defined as the following expected scale

$$p_{r,kj} = p(e_k^j(a_r)) = \sum_{n=1}^{N} \beta_{kj}^{n}(a_r) p(H_n) \quad (3.134)$$

where $p(H_n)$ is a real number defined to quantify H_n with $p(H_{n+1}) > p(H_n)$ if H_{n+1} is preferred to H_n. Details about how to define $p(H_n)$ can be found in [Yang and Singh 1994a][Yang and Sen 1994b].

Suppose $\lambda_k=[\lambda_k^1 \cdots \lambda_k^j \cdots \lambda_k^{L_k}]^T$ and λ_k^j represents the normalized relative weight of the factor e_k^j in evaluation of y_k where $0 \le \lambda_k^j \le 1$. λ_k^j may be assigned as follows.

Suppose ζ_k^j expresses the relative weight of the factor e_k^j in evaluation of y_k, and ζ_k can be articulated as a unit weight vector as follows

$$\zeta_k = [\zeta_k^1 \cdots \zeta_k^j \cdots \zeta_k^{L_k}]^T, \quad \sum_{j=1}^{L_k} \zeta_k^j = 1, \quad 0 \le \zeta_k^j \le 1 \quad (3.135)$$

ζ_k can be readily obtained using any well-known weight assignment method, such as the eigenvector method [Huang and Yong 1981] [Saaty 1988].

Let e_k^l be the most important factor in E_k, called the key factor, that is, $\zeta_k^l = \max_j \{\zeta_k^1, \cdots, \zeta_k^j, \cdots, \zeta_k^{L_k}\}$. Normalize ζ_k by $\overline{\zeta}_k^j = \zeta_k^j / \zeta_k^l$, $j=1, \cdots, L_k$. If for the key factor the following relation is true

$$m_{kl}^{n} = \alpha_k \beta_{kl}^{n}, \quad 0 < \alpha_k \leq 1 \tag{3.136}$$

then, λ_k^j (j=1 , $\cdots$, L_k) may be obtained by $\lambda_k^j = \alpha_k \overline{\zeta}_k^j$, j=1 , $\cdots$, L_k

α_k may be referred to as a priority coefficient representing the degree of significance of the role of the most importance factor in the evaluation of the attribute y_k. α_k is determined by satisfying the following relation

$$\prod_{j=1}^{L_k} \left[1 - \alpha_k \frac{\zeta_k^j}{\zeta_k^l} \right] \leq \delta, \qquad \delta \geq 0 \tag{3.137}$$

where δ is a small non-negative real number. It may be taken so that $1.0 \times 10^{-6} \leq \delta \leq 1.0 \times 10^{-2}$. Such an assignment means that if H_j is absolutely confirmed by all the factors E_k then the associated attribute will be confirmed to H_j to the extent of over $(1-\delta)\times 100$ percent. More details about the assignment of λ can be found in [Yang and Sen 1994b].

m_{kj}^n may then be calculated by

$$m_{kj}^{n}(a) = \lambda_k^j \beta_{kj}^{n}(a) \tag{3.138}$$

The above equation means that the fact that the state of e_k^j is absolutely evaluated to an evaluation grade H_n only supports to the extent of λ_k^j the hypothesis that the state of y_k is confirmed to H_n. It is obvious that $\sum_{n=1}^{N} m_{kj}^n \leq 1$. Suppose m_{kj}^H is the basic probability assignment to H, which is the remaining belief unassigned after commitment of belief to all H_n (n=1, $\cdots$, N), that is, $m_{kj}^H = 1 - \sum_{n=1}^{N} m_{kj}^n$. A basic probability assignment matrix $M(y_k/E_k)$ for evaluation of $y_k(a)$ through $E_k(a)$ may then be formulated by

$$M(y_k/E_k) = \begin{bmatrix} m_{k1}^1 & \cdots & m_{k1}^n & \cdots & m_{k1}^N & m_{k1}^H \\ \cdots & \cdots & \cdots & \cdots & \cdots & \cdots \\ m_{kj}^1 & \cdots & m_{kj}^n & \cdots & m_{kj}^N & m_{kj}^H \\ \cdots & \cdots & \cdots & \cdots & \cdots & \cdots \\ m_{kL_k}^1 & \cdots & m_{kL_k}^n & \cdots & m_{kL_k}^N & m_{kL_k}^H \end{bmatrix} \begin{matrix} \{e_k^1(a)\} \\ \cdots \\ \{e_k^j(a)\} \\ \cdots \\ \{e_k^{L_k}(a)\} \end{matrix} \tag{3.139}$$

Suppose Ψ is a subset of H, that is $\Psi \subseteq H$, and m_k^{Ψ} is an overall probability assignment to which the state of y_k at a_r is confirmed to Ψ by the factor set E_k, or $m_k^{\Psi} = m(\Psi/E_k(a_r))$. If m_k^{Ψ} for all $\Psi \subseteq H$ are generated from $M(y_k/E_k)$, then the state of y_k at a_r may be expressed by the following expectation

$$S(y_k(a_r)) = \{(m(\Psi/E_k(a_r)),\ \Psi),\ \text{for } all\ \Psi \subseteq H\} \tag{3.140}$$

The preference degree of $y_k(a_r)$, i.e. $p(y_k(a_r))$, is used to quantify $S(y_k(a))$ and may thus be defined as the following expected scale

$$p_{rk} = p(y_k(a_r)) = \sum_{\Psi \subseteq H} m(\Psi/E_k(a_r))p(\Psi) \tag{3.141}$$

where $p(\Psi)$ is the scale of Ψ and is defined as the average of $p(H_n)$ for all $H_n \subseteq \Psi$. A qualitative attribute quantified by a preference degree possesses the basic property of its marginal utilities being monotonous. In other words, for two designs, a_r and a_h, $S(y_k(a_r))$ is preferred to $S(y_k(a_h))$ if and only if $p_{rk} > p_{hk}$. Such quantification can thus form a rational basis for further decision analysis. A basic evidential reasoning algorithm is developed for generating m_k^{Ψ} from $M(y_k/E_k)$.

Define a subset $e_{I_k(i)}$ of E_k and a combined probability assignment $m_{I_k(i)}^{\Psi}$ as follows

$$e_{I_k(i)} = \{e_k^1 \cdots e_k^i\},\ \ 1 \le i \le L_k; \qquad m_{I_k(i)}^{\Psi} = m(\Psi/e_{I_k(i)}) \tag{3.142}$$

where $m(\Psi/e_{I_k(i)})$ is a combined probability assignment to Ψ confirmed by $e_{I_k(i)}$. Then, the algorithm can be stated as follows

$$\{H_j\}:\ m_{I_k(i+1)}^j = K_{I_k(i+1)}(m_{I_k(i)}^j m_{k,i+1}^j + m_{I_k(i)}^j m_{k,i+1}^H$$

$$+ m_{I_k(i)}^H m_{k,i+1}^j),\ \ j=1, \cdots, N \tag{3.143a}$$

$$\{H\}:\ m_{I_k(i+1)}^H = K_{I_k(i+1)} m_{I_k(i)}^H m_{k,i+1}^H \tag{3.143b}$$

$$K_{I_k(i+1)} = \left[1 - \sum_{\tau=1}^{N}\sum_{\substack{j=1\\ j\ne\tau}}^{N} m_{I_k(i)}^{\tau} m_{k,i+1}^j\right]^{-1} \tag{3.143c}$$

$$i=1, \cdots, L_k-1$$

c) Hierarchical evidence propagation

It can be proven from the algorithm that $m_{I_k(L_k)}^{\Psi}$ is the overall probability assignment to Ψ ($\subseteq H$) confirmed by E_k and $m_{I_k(L_k)}^{\Psi}=0$ for any $\Psi \subseteq H$ other than

$\Psi = H_j$ $(j=1, \cdots, N)$ and H, or

$$m_k^j = m(H_j/E_k) = m^j_{I_k(L_k)}, \quad j = 1, \cdots, N, \text{ and}$$

$$m_k^H = m(H/E_k) = m^H_{I_k(L_k)} \tag{3.144}$$

$$m(\Psi/E_k) = m^{\Psi}_{I_k(L_k)} = 0$$

$$\text{for } any\ \Psi \subseteq H\ but\ \Psi \neq H_j\ (j=1, \cdots, N) \text{ and } H \tag{3.145}$$

Consequently, the state of the kth attribute can be evaluated in terms of the evaluation grades defined in H as follows

$$S(y_k) = \{(m_k^j, \quad H_j), \ j=1, \cdots, N\} \tag{3.146}$$

that is, the kth attribute is evaluated to H_j with a degree of confidence of m_k^j, $j=1, \cdots, N$. Such an evaluation is generated by synthesizing the given judgements of the factors associated with the attribute.

The preference degree of $y_k(a_r)$, i.e. $p(y_k(a_r))$, is used to quantify $S(y_k(a))$ and may thus be obtained as the following expected scale

$$p_{rk} = p(y_k(a_r)) = \sum_{n=1}^{N} m^n_{I_k(L_k)} p(H_n) + m^H_{I_k(L_k)} p(H) \tag{3.147}$$

where $p(H) = \sum_{n=1}^{N} p(H_n)/N$, being the average value of $p(H_n)$.

In the above analysis, it is assumed that the evaluations of a factor are given. From Figure 3.15, it may be noted that a factor may only be directly evaluated at each of the three scenarios separately. Problems may then arise as to how to generate an overall evaluation of a factor by taking into account the three scenarios together. This can be readily done in the same way as before if the three scenarios are treated as imaginary sub-factors. The next section is meant to demonstrate such a hierarchical evidence reasoning process.

3.4.3 The Ship Choice Problem

a) The subjective assessments

Using the framework outlined in Section 3.4.2 requires the representation and processing of subjective data. For this ship choice problem the assessments of the nine basic factors are shown in tables 3.17 to 3.19.

Table 3.17 *Judgements of Each Ship on All Factors* for *Scenario* 1

Factors	Ship 1	Ship 2	Ship 3
Bale Utilization	G(0.5), A(0.5)	G(0.4), A(0.6)	G(0.4), A(0.6)
TEU Utilization	E(0.1), V(0.6), G(0.2)	E(0.2), V(0.6), G(0.1)	G(0.3), A(0.3), B(0.3)
DWT Utilization	V(0.2), G(0.5), A(0.3)	V(0.1), G(0.4), A(0.4)	V(0.1), G(0.5), A(0.2)
Maintenance & Repair	V(0.1), G(0.5), A(0.4)	G(0.2), A(0.6), B(0.1)	V(0.2), G(0.4), A(0.3)
Fuel Consumption	G(0.4), A(0.5), B(0.1)	A(0.4), B(0.5)	V(0.1), G(0.5), A(0.3)
Relative Freight Cost	E(0.3), V(0.3), G(0.4)	V(0.2), G(0.6)	E(0.2), V(0.2), G(0.5)
Resale Potential	V(0.1), G(0.4), A(0.4)	G(0.4), A(0.4)	G(0.5), A(0.5)
Frequency of Service	V(0.4), G(0.6)	V(0.3), G(0.7)	V(0.3), G(0.3), A(0.3)
Cargo Expedition	E(0.5), V(0.2), G(0.2)	E(0.5), V(0.5)	E(0.1), V(0.3), G(0.4)

Table 3.18 *Judgements of Each Ship on All Factors* for *Scenario* 2

Factors	Ship 1	Ship 2	Ship 3
Bale Utilization	G(0.5), A(0.4), B(0.1)	G(0.2), A(0.4), B(0.3)	G(0.4), A(0.6), B(0.2)
TEU Utilization	G(0.3), A(0.5), B(0.1)	G(0.2), A(0.4), B(0.4)	G(0.3), A(0.3), B(0.3)
DWT Utilization	G(0.5), A(0.4), B(0.1)	G(0.3), A(0.4), B(0.3)	G(0.5), A(0.5)
Maintenance & Repair	V(0.1), G(0.5), A(0.4)	G(0.4), A(0.5)	V(0.2), G(0.4), A(0.3)
Fuel Consumption	G(0.6), A(0.4)	A(0.6), B(0.4)	G(0.7), A(0.3)
Relative Freight Cost	E(0.3), V(0.4), G(0.3)	G(0.3), A(0.5)	E(0.3), V(0.3), G(0.3)
Resale Potential	G(0.4), A(0.6)	G(0.3), A(0.7)	G(0.5), A(0.5)
Frequency of Service	V(0.2), G(0.6), A(0.1)	V(0.1), G(0.4), A(0.5)	V(0.2), G(0.5), A(0.2)
Cargo Expedition	V(0.6), G(0.3), A(0.1)	E(0.1), V(0.4), G(0.4)	V(0.4), G(0.5), A(0.1)

The assessment grades are given by

$$\begin{aligned} H &= \{H_1 \quad H_2 \quad H_3 \quad H_4 \quad H_5 \quad H_6\} \\ &= \{Poor \quad Below\ Average \quad Average \\ &\quad Good \quad Very\ Good \quad Excellent\} \end{aligned} \tag{3.148}$$

and these grades are quantified by

$$\begin{aligned} p\{H\} &= (p(H_1) \quad p(H_2) \quad p(H_3) \quad p(H_4) \quad p(H_5) \quad p(H_6)) \\ &= (-1 \quad -0.6 \quad -0.2 \quad 0.2 \quad 0.6 \quad 1) \end{aligned} \tag{3.149}$$

The subjective assessments over the attributes and factors show that the notional operator favours the modest sized Ship 1 design over the larger and smaller

Table 3.19 *Judgements of Each Ship on All Factors* for *Scenario* 3

Factors	Ship 1	Ship 2	Ship 3
Bale Utilization	G(0.3), A(0.7)	G(0.4), A(0.6)	G(0.4), A(0.5), B(0.1)
TEU Utilization	V(0.2), G(0.6), A(0.1)	V(0.2), G(0.6), A(0.2)	G(0.6), A(0.4)
DWT Utilization	G(0.4), A(0.4), B(0.1)	G(0.3), A(0.4), B(0.2)	G(0.4), A(0.5)
Maintenance & Repair	G(0.4), A(0.6)	G(0.3), A(0.6)	G(0.5), A(0.5)
Fuel Consumption	G(0.4), A(0.5)	G(0.4), A(0.4), B(0.1)	G(0.4), A(0.6)
Relative Freight Cost	E(0.3), V(0.2), G(0.5)	G(0.4), A(0.5)	E(0.1), V(0.3), G(0.3) A(0.2)
Resale Potential	G(0.5), A(0.4)	G(0.7), A(0.3)	G(0.4), A(0.6)
Frequency of Service	V(0.3), G(0.6)	V(0.3), G(0.5), A(0.2)	V(0.3), G(0.6)
Cargo Expedition	V(0.6), G(0.4)	V(0.7), G(0.3)	V(0.4), G(0.6)

options. It will be noted that Ship 2 is not greatly favoured because of its comparatively large size and speed. This would mean that such a vessel would offer a lower frequency of service even though each unit of cargo, once on board, will be experiencing a speedier passage. Its main advantage will be in Scenario 1 when most of the cargo is containerised. Ship 3 is not favoured as it is perceived by the operator to be similar to Ship 1 but with a significantly lower TEU capacity. It is also slightly slower and this has a modest detrimental effect on service factors.

The relative freight cost is taken as the measure of merit because the influence of ship features on average freight revenues that can be expected is to be subjectively taken into account. The initial weightings associated with the factors and the scenarios are shown by value system (or V-S) 1 in Table 3.20, 3.21 and 3.22. The three additional weightings are shown by value systems 2 and 3 in Table 3.20 for the attributes and by value system 2 in Table 3.21 for the scenarios.

Table 3.20 *Weight Assignment* for *Attributes*

Atribute	Value System 1	Value System 2	Value System 3
Price	10	10	10
Speed	5	5	10
Cargo	7.5	10	5
Economic	10	5	10
Service	7.5	10	10

Table 3.21 *Weight Assignment* for *Scenarios*

Scenarios	Scenario 1	Scenario 2	Scenario 3
Value System 1	0.33	0.33	0.33
Value System 2	0.60	0.20	0.20

Table 3.22 *Weights of Factors* for *Each Attribute*

Attributes	Factors	Weights
Cargo	Bale Utilization	1.0
	TEU Utilization	1.0
	DWT Utilization	1.0
Econimoc	Maintenance & Repair	0.5
	Relative Fuel Consumption	0.5
	Relative Freight Cost	1.0
	Resale Potential	0.5
Service	Frequency of Service	1.0
	Cargo Expedition	1.0

b) Results analysis

The results of analysis are presented in Tables 3.23 to 3.26. The rankings of the three ships as shown in Table 3.26 are obtained using the CODASID method based on Table 3.25. It can be clearly seen that on the basis of all of the priority orderings, Ship 1 comes out on top. It is instructive to examine the results in some detail.

For the basis weighting Ship 1 comes out as the best for all the top level factors except speed and service friendliness, where Ship 2 is clearly the best. However, these advantages of Ship 2 are not enough to compensate for the other factors. Its cargo characteristics and unfavourable cost features keep it in the third position. It is only when speed and cargo factors are given relative importance that Ship 2 comes in as the second alternative, ahead of Ship 3.

When the scenarios are weighted in favour of the containerised route of Scenario 1 then Ship 2 performs considerably better. Its service friendliness is particularly valued and its cargo characteristics improve in relative terms. This leads to Ship 2 being consistently evaluated to be preferable to Ship 3.

The results indicate that for the hypothetical owner in question, Ship 1 is the best alternative but there are circumstances under which he ought to value the special advantages of Ship 2. These results are consistent with those obtained earlier (Buxton 1989). This is not surprising as the subjective judgements were informed by the results of earlier analysis, but it does give confidence in the proposed methodology.

Table 3.23 *Judgements of Each Ship* for *All Scenarios*

Factors	Ship 1	Ship 2	Ship 3
Bale Utilization	G(0.3854), A(0.5810) B(0.0090)	G(0.2749), A(0.6651) B(0.0267)	G(0.3711), A(0.5656) B(0.0364)
TEU Utilization	E(0.0211), V(0.2361) G(0.4883), A(0.1545) B(0.0211)	E(0.0340), V(0.2226) G(0.3803), A(0.2067) B(0.0927)	G(0.4775), A(0.3566) B(0.1148)
DWT Utilization	V(0.0270), G(0.4938) A(0.4151), B(0.0270)	V(0.0135), G(0.3435) A(0.4807), B(0.1118)	V(0.0103), G(0.5342) A(0.4069)
Maintenance & Repair	V(0.0223), G(0.4801) A(0.4717)	G(0.2267), A(0.7123) B(0.0128)	V(0.0627), G(0.5130) A(0.3779)
Fuel Consumption	G(0.4785), A(0.4785) B(0.0115)	G(0.0510), A(0.6022) B(0.2991)	V(0.0084), G(0.6121) A(0.3479)
Relative Freight Cost	E(0.2672), V(0.2581) G(0.4420)	V(0.0355), G(0.5601) A(0.3200)	E(0.1715), V(0.2726) G(0.4507), A(0.0366)
Resale Potential	V(0.0108), G(0.4389) A(0.5099)	G(0.4810), A(0.4810)	G(0.4226), A(0.5546)
Frequency of Service	V(0.2096), G(0.7446) A(0.0097)	V(0.1487), G(0.7055) A(0.1159)	V(0.2383), G(0.5960) A(0.1057)
Cargo Expedition	E(0.0553), V(0.6030) G(0.2852), A(0.0151)	E(0.0968), V(0.7406) G(0.1251)	E(0.0081), V(0.3406) G(0.5988), A(0.0140)

Table 3.24 *Judgements of Each Ship on Every Attribute*

Attributes	Ship 1	Ship 2	Ship 3
Cargo	E(0.0025), V(0.0338) G(0.5451), A(0.3692) B(0.0077)	E(0.0047), V(0.0337) G(0.3370), A(0.5384) B(0.0399)	E(0.0000), V(0.0011) G(0.4899), A(0.4544) B(0.0186)
Economic	E(0.1209), V(0.1214) G(0.6258), A(0.0851) B(0.0005)	E(0.0000), V(0.0128) G(0.3937), A(0.5234) B(0.0147)	E(0.0679), V(0.1178) G(0.6392), A(0.1203)
Service	E(0.0125), V(0.3805) G(0.5664), A(0.0059)	E(0.0273), V(0.4740) G(0.4222), A(0.0345)	E(0.0018), V(0.2274) G(0.7121), A(0.0256)

Table 3.25 *Decision Matrix* for *The Ship Choice Problem*

	Price	Speed	Cargo	Economic	Service
Ship 1	16.66	15.0	0.053327	0.301617	0.352963
Ship 2	25.00	17.6	-0.039347	-0.027092	0.389262
Ship 3	16.66	14.3	-0.003371	0.242373	0.275511

Table 3.26 *Rankings of The Three Ships* for *Various Value Systems*

Scenarios	Value System 1			Value System 2		
Attributes	V-S 1	V-S 2	V-S 3	V-S 1	V-S 2	V-S 3
Ship 1	1	1	1	1	1	1
Ship 2	3	3	2	2	2	2
Ship 3	2	2	3	3	3	3

3.5 Concluding Comments

The aim of this chapter has been to present a range of MADM techniques along with their principal characteristics with a view to facilitating their formal application in decision making. Selection in the presence of multiple attributes is the commonest form of decision problem that surfaces naturally in real life. We all engage in it in one form or another. Hence it is important that the use of appropriate methodology is pursued in the context of an adequate understanding of the associated processing logic.

The generation of technical alternatives is often a very expensive operation, involving an estimation of performance attributes of various kinds. It is right and proper, therefore, that the evaluation of the alternatives receives equal care so that the best use is made of the information available. This should help to make the selection exercise more rational. Moreover, given the intimate relationship between MADM and MODM, it is important for more informed design synthesis to understand the underlying trade-offs in the criteria domain and their relationship with trade-offs in the design variable domain.

4
Multiple Objective Decision Making

4.1 Multiobjective Optimisation and Method Classification

4.1.1 Multiobjective Optimisation and Utility Functions

A multiobjective optimisation problem may generally be represented as the following vector mathematical programming problem

$$MOP \begin{cases} optimise & F(X) = \{f_1(X) \cdots f_i(X) \cdots f_k(X)\} \\ subject\ to & X \in \Omega \end{cases} \tag{4.1}$$

$$\Omega = \left\{ X \;\middle|\; \begin{array}{ll} g_i(X) \le 0 & i=1, \cdots, m_1 \\ h_j(X) = 0 & j=1, \cdots, m_2 \\ X = [x_1 \cdots x_n]^T & \end{array} \right\}$$

where x_i is a decision variable, X denotes a solution, $f_i(X)$ is generally a nonlinear objective function, respectively, and $g_i(X)$ and $h_j(X)$ are nonlinear inequality and equality constraint functions. These objectives are usually incommensurate and in conflict with one another. Therefore there normally exist infinite number of efficient (noninferior, non-dominated or Pareto-optimal) solutions in the MOP. The problem is how to search for a best compromise solution with these multiple objectives being considered simultaneously.

The best compromise solution should be a solution that can best satisfy the DM's preferences. If the DM's preferences are modeled by a utility function aggregating all objective functions into one criterion, denoted by

$$u(F(X)) = u(f_1(X) \cdots f_i(X) \cdots f_k(X)) \quad \text{for } X \in \Omega \tag{4.2}$$

then the best compromise solution may be defined as a solution that maximises the utility function $u(F(X))$ for all $X \in \Omega$.

For convenience of discussion, we introduce the following definitions about decision space, objective space, efficient solution, weakly efficient solution and efficient frontier.

Definition 4.1: The feasible space Ω is the set of all feasible solutions that satisfy the constraints, as defined in problem (4.1).

Definition 4.2: The feasible objective space Λ is the image by F of the feasible decision space Ω, i.e.

$$\Lambda = \{F \mid F = F(X),\ X \in \Omega\} \tag{4.3}$$

Definition 4.3: X^t is called an efficient (noninferior, nondominated or Pareto-optimal) solution of problem (4.1) if there does not exist any $X \in \Omega$, $X \neq X^t$, so that $F(X) \leq F(X^t)$ and $F(X) \neq F(X^t)$, where f_i $(i=1, \cdots, k)$ are assumed for minimisation.

Definition 4.4: X^t is called a weakly efficient solution of problem (4.1) if there does not exist any $X \in \Omega$, $X \neq X^t$, so that $F(X) < F(X^t)$, where f_i $(i=1, \cdots, k)$ are assumed for minimisation.

Definition 4.5: The collection of all efficient solutions of problem (4.1) is called the efficient set. The image of the efficient set by F is referred to as the efficient frontier or tradeoff surface.

Figure 4.1 illustrates the feasible decision space with two decision variables (x_1 and x_2) and the feasible objective space. Note that each point in the feasible objective space Λ is in correspondence with one or multiple points in the feasible decision space Ω, depending upon the characteristics of the mapping (objective) functions f_i $(i=1, \cdots, k)$.

Figure 4.2 illustrates efficient solutions, weakly efficient solutions and efficient frontier, where f_1 and f_2 are assumed for minimisation. Note that all points on the curve between points F^1 and F^2 are weakly efficient solutions (F^2 itself is an efficient solution). Also note that F^3 is only a weakly efficient solution as $f_1^4 < f_1^3$ and $f_2^4 = f_2^3$.

From Figure 4.2, one can find that in a multiobjective optimisation problem there are normally infinite number of efficient solutions due to the conflict between objectives. In other words, there is normally not a single solution which could optimise all objectives simultaneously. Therefore, multiple objective decision making usually comprises two main steps, generation of efficient solutions and

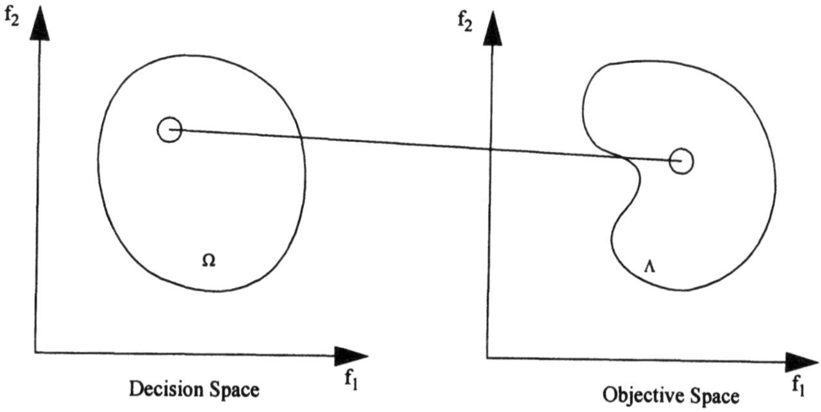

Figure 4.1 *The Mapping of Decision Space into Objective Space*

identification of the best compromise solution which must be an efficient solution. In the past two decades or so, a large number of MODM methods have been developed.

The additive weighting method is one of the simplest and most commonly used techniques. In this scheme a weight vector is chosen. The chosen weights may be used as parameters which can be regulated or as relative weights representing the DM's preferences. In the latter case, a utility function is defined as the weighted sum of objective functions. A scalar objective function problem may thus be defined for the simple weighting method as follows

$$\begin{aligned} \min \;\; & u = W^T F(X) \\ s.t. \;\; & X \in \Omega \qquad W = [w_1 \cdots w_i \cdots w_k]^T \end{aligned} \tag{4.4}$$

where w_i is the weighting factor of objective i.

In objective space, problem (4.4) can be rewritten as follows

$$\begin{aligned} \min \;\; & u = W^T F \\ s.t. \;\; & F \in \Lambda \end{aligned} \tag{4.5}$$

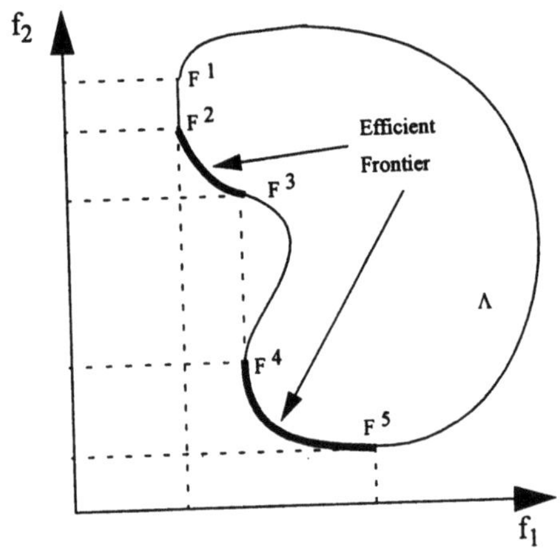

Figure 4.2 *(Weakly) Efficient Solutions*

where Λ is defined by equation (4.3).

For a given utility, for example u_0, one can find that the equation $u_0 = W^T F$ is a hyperplane in the objective space with normal W. Thus, solving problem (4.5) is equivalent to finding the smallest value of u_0 for which there exists an $F \in \Lambda$ with $W^T F = u_0$. Clearly this is a point where the hyperplane with normal W is just tangential to Λ. Such a point is usually an (weakly) efficient solution. Given different values of W, different (weakly) efficient solutions may be found. Figure 4.3 illustrates this situation. Furthermore, all (weakly) efficient solutions may be found using the weighting method by regulating W if the feasible objective space Λ is convex.

If Λ is non-convex, however, the simple weighting method may not be capable of identifying all efficient solutions, as shown by Figure 4.4, because not every efficient solution has a supporting hyperplane. In Figure 4.4, each point on the curve between points F^1 and F^2 is an efficient solution but none of them could

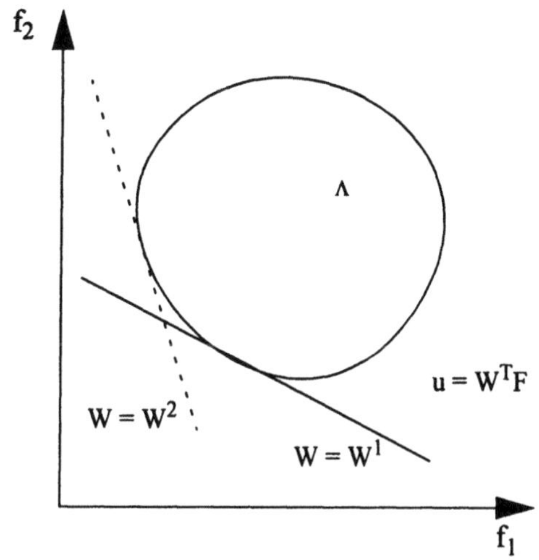

Figure 4.3 *The Simple Weighting Method*

be identified using the simple weighting method no matter what weight or optimisation technique is used.

If in problem (4.4) W is not used as weighting parameters but treated as relative weights representing the DM's preferences, a linear additive utility function is actually adopted. In utility theory, such a function necessarily assumes that the marginal utility function of an objective function is linear, such that any objectives can be evaluated independently of other objectives and bad values of some objectives can be directly offset without limit by good values of other objectives. However, these assumptions may not always be acceptable in real-life problems.

A linear marginal utility function means, for example, that the change of utility for each unit change of an objective is always the same regardless of the base values of the objective. If salary is an objective, for example, a linear utility function of salary necessarily implies that the same pay rise (say 1000 pounds) would be equally preferred no matter what the base salary may be (say 5000 or 20000 pounds), as shown by Figure 4.5. However, this may not be the case. When the base salary is only 5000 pounds, for example, one may be quite happy with a 1000 pound pay rise but may not be equally impressed by the same pay

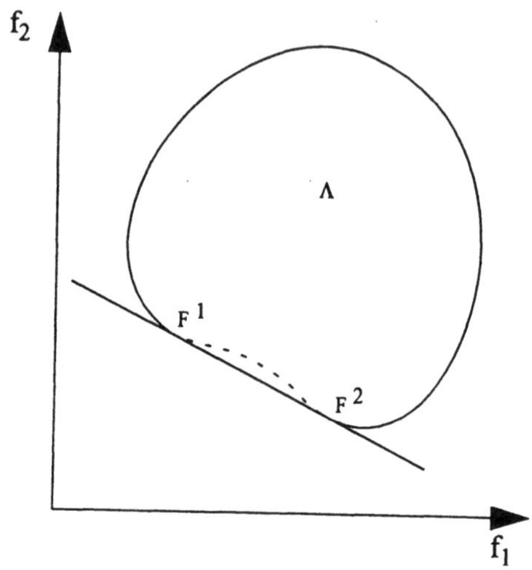

Figure 4.4 *The Dashed Portion of Efficient Frontier Cannot Be Discovered Using The Simple Weighting Method*

rise when the base salary is 20000 pounds. This means that the utility function of salary should be nonlinear, as illustrated by Figure 4.6.

4.1.2 Classification of MODM Methods

Just as selection (MADM) methods of Chapter 3 depend on prioritised attributes, synthesis or configuration of alternatives (or MODM) depend on prioritised objectives. Classification of MODM techniques provides a framework on which advice can be based as to the aptness of a method to a specific problem situation. For generic classification of MODM techniques, several schemata have been proposed. For instance, Teghem et al. used six criteria to classify MODM methods [Teghem et al. 1989]. Evans presents a schema to classify MODM methods according to the timing of the elicitation of preference information [Evans 1984].

As far as we know, however, there has been no single, widely adopted classification for MODM methods. This section is not intended to suggest a generic classification schema either but to propose a specific classification framework for application of MCDM methods in engineering design. Such a

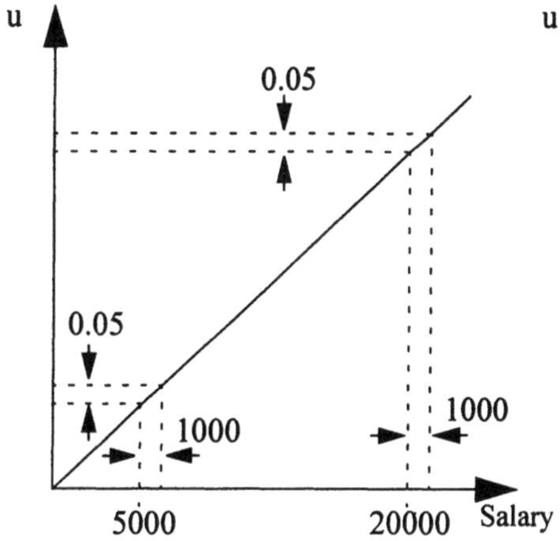

Figure 4.5 *Linear Utility Function*

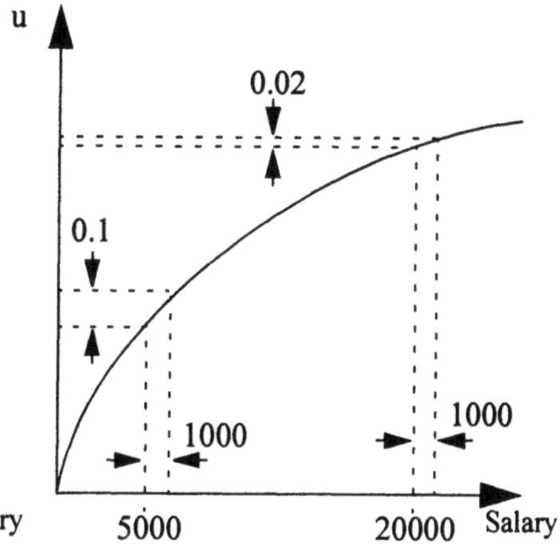

Figure 4.6 *Nonlinear Utility Function*

framework is shown in Figure 4.7. The methods listed in Figure 4.7 are only some of the MODM methods which have been developed in the past two decades. Some of the listed methods may be selected for the development of decision support systems for application in engineering design. The rules for selecting an appropriate MODM method are discussed as follows.

The classification of MODM methods, as shown in Figure 4.7, is mainly based on the types of preference information and the timing for eliciting preference information. There are generally three timing options for eliciting preferences: *a posteriori* articulation, progressive articulation and *a priori* articulation. There are four major types of preference information, ordinal information, cardinal information, implicit and explicit trade-off information.

In addition to the types and the timing of the elicitation of preference information, a choice rule for selecting a MODM method may take into account the decision rule or the computational procedure of the method. To choose the minimax (or ideal point) method, for example, the choice rule may be constructed as follows

> **if** *the overall preference information is given beforehand in the form of a pre-defined ideal design and subjective weights representing the relative importance of objectives, and*
>
> **if** *the decision rule is to choose a design which is closest to the pre-defined ideal design,*

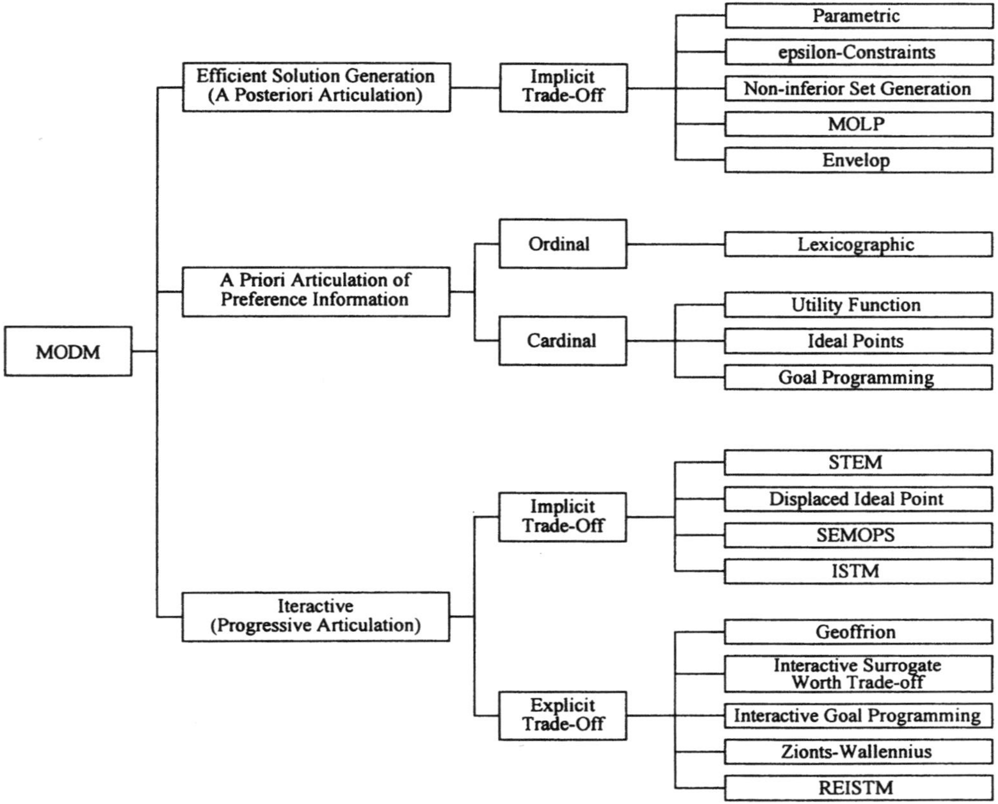

Figure 4.7 *Classification of MODM Methods*

then *the ideal point method is suggested.*

The choice rule for the ISTM method is listed below

> **if** *preference information can only be provided progressively as alternative designs are generated,*
>
> **if** *preference information can only be given in terms of implicit tradeoffs between objectives, and*
>
> **if** *the designer can identify three distinct subsets of objectives, that is the subset of objectives for improvement, the subset for maintenance and the subset for sacrifice, and he can also indicate the amount to which an objective in the third subset may be sacrificed from the current levels,*
>
> **then** *the ISTM method is suggested.*

A decision tree for selecting a MODM method is shown in Figure 4.8. Some of the MODM methods in the decision tree have been chosen for the development of a decision support system: in particular, the minimax method, goal programming, the interactive step trade-off method (ISTM) and Geoffrion's method.

A MODM method may be selected for this purpose if it can deal with nonlinear problems as most multiobjective design synthesis problems are strongly nonlinear. This explains why those MODM methods which can only deal with linear problems should not be selected for the development of a decision support system for application in engineering design.

Secondly, at preliminary design stage of a new large engineering product, a designer may not be able to provide *a priori* preference information. From this point of view, interactive MODM methods are the most appropriate for specific application in engineering design. Such techniques allow the solution to progress towards a preferred solution through a learning-oriented or search-oriented interactive procedure. This mirrors the common engineering design process of evolving design and adaptive design. The ISTM method and Geoffrion's method may be selected as a typical learning-oriented interactive method and a typical search-oriented interactive method, respectively. The ideal point method may also be used as an interactive technique by regulating the weighting parameters in a systematic way.

Furthermore, the ideal point method and the ISTM method may also be used to generate the whole set of efficient solutions of a multiobjective problem. If a designer is able to provide *a priori* articulation of preferences, however, the goal programming and the ideal point method may be appreciated because they use simple logic and require much less computational time than the interactive methods. After discussion of some common optimality conditions, the next a few sections desribe the basic optimisation methods chosen for the design decision support system as well as the new MODM method that has been developed. The methods are also examined in the context of application.

4.2 Techniques for Single-Objective Optimisation

4.2.1 Optimality Conditions

As mentioned in the last section, one of the main steps in multiobjective optimisation is the generation of efficient solutions. An efficient solution can be obtained by constructing and solving an auxiliary single-objective optimisation problem, as generally defined by

$$\begin{aligned} \min \ & f(X) \\ s.t. \ & X \in \Omega \end{aligned} \tag{4.6}$$

where Ω is defined as in problem (4.1) and $f(X)$ is a differentiable function.

To obtain optimality conditions, slack variables $s_i \geq 0$ are introduced for inequality constraints and problem (4.6) can be rewritten as follows

$$\begin{aligned} \min \ & f(X) \\ s.t. \ & g_i(X) - s_i = 0 \qquad i = 1, \cdots, m_1 \\ & h_j(X) = 0 \qquad j = 1, \cdots, m_2 \\ & s_i \geq 0 \end{aligned} \tag{4.7}$$

The Lagrangian function for problem (4.7) is then given by

$$L(X, s, \lambda, \mu) = f(X) + \sum_{i=1}^{m_1} \lambda_i (g_i(X) - s_i) + \sum_{j=1}^{m_2} \mu_j h_j(X) \tag{4.8}$$

Suppose X^0 is a solution of problem (4.6) or $X^0 \in \Omega$. Then, a necessary condition for X^0 to be a local optimum of problem (4.6) is that there exist real numbers $\lambda_i \geq 0$ for all i=1, $\cdots$, m_1 and μ_j for all j=1, $\cdots$, m_2 (μ_j not sign-restricted) such that

$$\begin{aligned} & \nabla f(X^0) + \sum_{i=1}^{m_1} \lambda_i \nabla g_i(X^0) + \sum_{j=1}^{m_2} \mu_j \nabla h_j(X^0) = 0 \\ & \lambda_i g_i(X^0) = 0 \ \text{ for all } i=1, \cdots, m_1 \end{aligned} \tag{4.9}$$

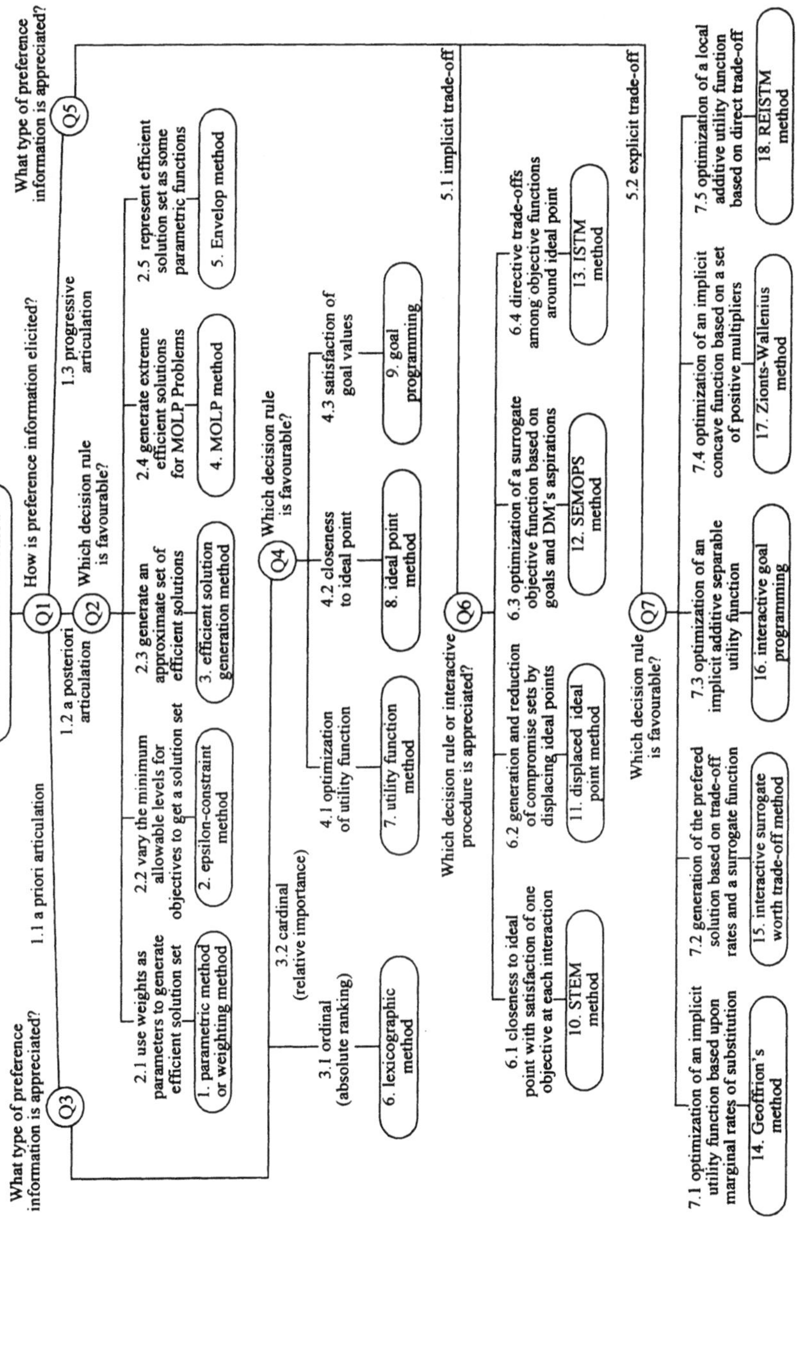

Figure 4.8 *Decision Tree for Selecting a MODM Method*

Note that $\lambda_i = 0$ if $g_i(X^0) < 0$.

Conditions (4.9) are referred to as Kuhn-Tucker optimality conditions. Generally, it is a necessary condition for X^0 ($\in \Omega$) to be a local optimum of problem (4.6). If in problem (4.6) $f(X)$ is concave and Ω is a convex set, then the Kuhn-Tucker conditions are also sufficient conditions.

In conditions (4.9), λ_i $(i=1, \cdots, m_1)$ are Kuhn-Tucker multipliers. It is important to note that the Kuhn-Tucker vector $\lambda = [\lambda_1 \cdots \lambda_k]^T$ is unique when the gradients of the saturated constraints are linearly independent. In this case, X^0 is called a regular point. Figure 4.9 illustrates the Kuhn-Tucker conditions by a two-dimensional example. In this example, X^0 is a regular point as the gradients $\nabla g_1(X^0)$ and $\nabla g_2(X^0)$ of the two constraints g_1 and g_2 saturated at X^0 are linearly independent.

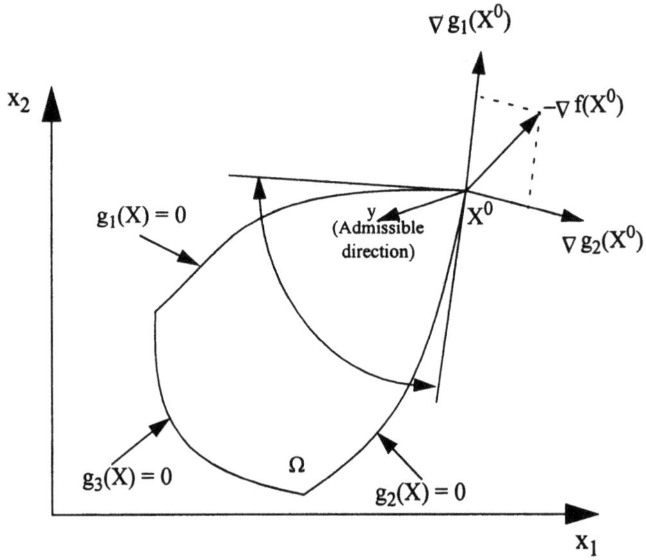

Figure 4.9 *The Kuhn-Tucker Optimization Conditions*

At X^0, the negative gradient $-\nabla f(X^0)$ of the objective function $f(X)$ is the linear combination of $\nabla g_1(X^0)$ and $\nabla g_2(X^0)$. Thus, an infinitesimal move along the direction $-\nabla f(X^0)$ will lead outside of the feasible decision space Ω. In other words, X^0 is at least a local minimum of $f(X)$ as no other feasible solution in a neighbourhood of X^0 could have a lower value of $f(X)$. Note that at X^0,

$$\begin{aligned} &g_1(X^0) = 0, \quad g_2(X^0) = 0 \\ &-\nabla f(X^0) = \lambda_1 \nabla g_1(X^0) + \lambda_2 \nabla g_2(X^0) \end{aligned} \tag{4.10}$$

with λ_1, $\lambda_2 > 0$. Thus, the Kuhn-Tucker conditions are satisfied at X^0.

The Kuhn-Tucker conditions can be extended to multiobjective optimisation problems. Suppose X^0 satisfies all the constraints of problem (4.1) and it is a regular point. If X^0 is also an efficient solution of problem (4.1), then there exist real numbers $\beta_t \geq 0$ for all $t=1, \cdots, k$, $\lambda_i \geq 0$ for all $i=1, \cdots, m_1$ and μ_j for all $j=1, \cdots, m_2$ (μ_j not sign-restricted) such that

$$\begin{aligned} &\sum_{t=1}^{k} \beta_t \nabla f_t(X^0) + \sum_{i=1}^{m_1} \lambda_i \nabla g_i(X^0) + \sum_{j=1}^{m_2} \mu_j \nabla h_j(X^0) = 0 \\ &\lambda_i g_i(X^0) = 0 \quad \text{for } all \ i=1, \cdots, m_1 \\ &X^0 \in \Omega \end{aligned} \tag{4.11}$$

Note that the above conditions are only necessary conditions for X^0 to be an efficient solution of problem (4.1).

4.2.2 Sequential Linear Programming

To obtain an optimal solution of problem (4.6), a number of methods can be used. These methods may be generally classified into two classes, primal methods and dual methods. Sequential linear programming (SLP) is a primal method, which has been developed due to the success of the Simplex method for solving linear programming. The guiding principle for this method consists of replacing the solution of a nonlinear problem by the solution of a sequence of linear ones which approximate the given problem in a sense that each nonlinear constraint (or objective) function is approximated by its first order Taylor expansion.

To solve a nonlinear optimisation problem, SLP first linearises the nonlinear objective function and the nonlinear constraint functions at a selected basic point. Around this point, a search space is set up based on the step sizes of variables. If the intersection of the linearised feasible space and the established search space is empty, the search space is then expanded by increasing the step sizes. If the intersection is non-empty, SLP will search for an optimal solution for the linearised objective function within a modified feasible space defined by the intersection. The obtained optimal solution is then used as a new basic point for re-linearisation of the nonlinear functions. The process is repeated until the real optimum of the original nonlinear problem is found.

This method does not converge in general without a proper regulation of the step sizes of variables, as can be easily observed from an example (Figure 4.10) [Minoux 1986]. The region of this example is a convex polyhedron and the lines of constant objective values are concentric circles. $\bar{f}(X)$ is the tangent line of $f(X)$ at point X. Taking X^0 as the starting basic point, we obtain X^1 and X^2 successively. Then, these last two points (X^1 and X^2) alternate indefinitely. This phenomenon is referred to as oscillation.

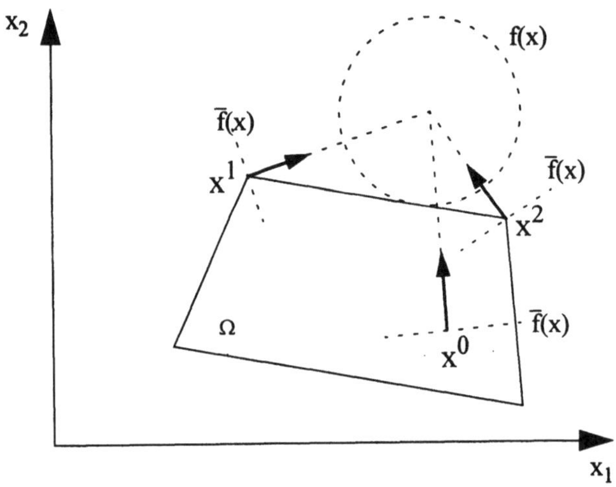

Figure 4.10 *Oscillation of Sequential Linear Programming*

To overcome the difficulty of oscillation, various techniques can be used [Minoux 1986]. One of such techniques is to regulate the step sizes of variables on line. Once oscillation is detected, for example, the step sizes can be reduced by a certain amount.

In an example as shown by Figure 4.11, for instance, a single variable function $y(x)$ is for maximisation. The nonlinear function is linearised at an initial point x^0. A search space (interval) around x^0 is established by the step size t^0. The maximum of the linearised function (the tangent line at x^0) is sought within the established search space, which leads to x^1. Repeating the same process with the fixed step size t^0, we can find x^2, x^3 and x^4 successively. Since $x^4 = x^2$, the first oscillation occurs. If the step size is reduced so that $t^1 = t^0/2$, for example, then we can find x^5 and x^6 successively. As $x^6 = x^4$, the step size may be further reduced with for example $t^2 = t^1/2$. We then get x^7 which is quite close to the maximum of $y(x)$.

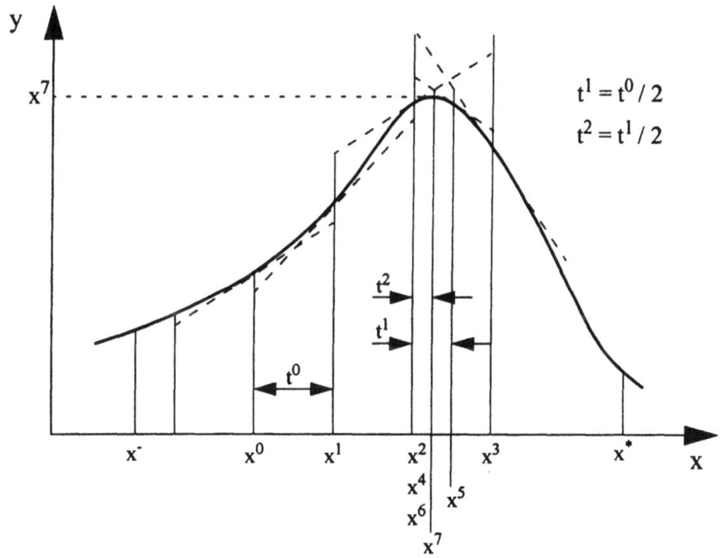

Figure 4.11 *Illustration of sequential linear programming with on-line step change*

It should be noted that for multivariable problems oscillation may occur between the current solution and a solution of a previously linearised problem (not the most recent one). It is therefore necessary to record previously generated solutions for checking such oscillation. How many previous solutions should be recorded depends on the complexity of a problem in question.

The computational steps of SLP may be summarised below.

Step 1: Define a nonlinear single-objective optimisation problem as follows

$$\begin{aligned} \min \quad & f(X) \\ s.t. \quad & x \in \Omega \end{aligned} \tag{4.12}$$

where Ω is defined in (4.1).

Step 2: Select an initial basic point X^0 which may be either feasible or infeasible. Initialise the step sizes of design variables denoted as Δx_i^0 $(i=1, \cdots, n)$. Let $t=0$.

Step 3: Linearise the nonlinear objective function $f(X)$ and nonlinear constraint functions $g_i(X)$ and $h_j(X)$ $(i=1, \cdots, m_1$ and $j=1, \cdots, m_2)$, at the point X^t by means of the first order Taylor expansions of these functions, that is

$$\bar{f}(X) = f(X^t) + \frac{\partial f(X^t)}{\partial X}(X - X^t) \tag{4.13}$$

$$\bar{g}_i(X) = g_i(X^t) + \frac{\partial g_i(X^t)}{\partial X}(X - X^t) \tag{4.14}$$

$$\bar{h}_j(X) = h_j(X^t) + \frac{\partial h_j(X^t)}{\partial X}(X - X^t) \tag{4.15}$$

Step 4: Using the current basic point and step sizes, define a search space S^t as follows

$$S^t = \left\{ X \;\middle|\; \begin{array}{ll} x_i^t - \Delta x_i^t \le x_i \le x_i^t + \Delta x_i^t, & i=1, \cdots, n \\ X = [x_1, x_2, \cdots, x_n]^T & \end{array} \right\} \tag{4.16}$$

Step 5: Formulate and solve the following linearised problem

$$\begin{aligned} \min \quad & \bar{f}(X) \\ s.t. \quad & S^t \cap \Omega^t \end{aligned} \tag{4.17}$$

where

$$\Omega^t = \left\{ X \left| \begin{array}{ll} \bar{g}_i(X) \leq 0 & i = 1, \ldots, m_1 \\ \bar{h}_j(X) = 0 & j = 1, \cdots, m_2 \end{array} \right. \right\}$$

Step 6: If Ω^t is empty, select a new basic point X^t and then go to step 3.

Step 7: If $S^t \cap \Omega^t$ is empty, increase the step sizes Δx_i by certain amount such as ten percent, that is $\Delta x_i^t = 1.1 \Delta x_i^t$. Then go to step 4.

Step 8: Solve linearised problem (4.17) using the Simplex method. If the optimal solution of (4.17) is X^{t+1} and it is equal to one generated before, oscillation is present which means that the optimum is an internal point. Then reduce the step sizes for all design variables by certain significant amount such as fifty percent, so that $\Delta x_i^t = 0.5 \Delta x_i^t$. Let $t = t + 1$ and go on.

Step 9: If X^{t+1} is infeasible for the original nonlinear problem, use X^{t+1} as a new basic point, let $t = t + 1$, and then go to step 3.

Step 10: If X^{t+1} is also feasible for the original nonlinear problem (4.12), then X^{t+1} will be taken as the optimal solution of (4.12) and the iterations stop only if either

1> the step sizes of all variables have been reduced to values below agreed thresholds, or

2> both design variables and objective values have not significantly changed in the last iteration.

Otherwise, use X^{t+1} as a new basic point, let $t = t + 1$, and then go to step 3.

For strongly nonlinear problems, Ω^t might not always be non-empty for any basic points while the original nonlinear problem may have feasible solutions. In such cases, the choice of the initial basic point X^0 and the initial step sizes ΔX^0 becomes very important. Unfortunately, SLP doesn't provide any systematic way to select X^0 and ΔX^0 which could guarantee that the Ω^t is non-empty.

4.2.3 Penalty Methods

Apart from the primal methods, dual methods form another class of techniques for nonlinear constrained optimisation. Penalty methods are widely used dual methods. They are simple and efficient for quickly obtaining good approximate solutions of a problem as well as good approximations to the optimal Kuhn-Tucker multipliers.

The basic idea of the penalty methods is to replace problem (4.6) by the following unconstrained optimisation problem (penalised problem)

$$P(\gamma_1, \gamma_2) \quad \begin{cases} \textit{min.} & \Gamma(X, \gamma_1, \gamma_2) = f(X) + \gamma_1 H_1(X) + \gamma_2 H_2(X) \\ \textit{s.t.} & X \in \Omega \end{cases} \tag{4.18}$$

where Γ is the pseudo-objective function, H_1 the penalty function for the equality constraints $h_j(X) = 0$, $j=1, \cdots, m_1$, H_2 the penalty function for the inequality constraints $g_i(X) \leq 0$, $i=1, \cdots, m_2$, and γ_1 and γ_2 are the penalty coefficients.

The penalty functions H_1 and H_2 may take different forms. To apply such techniques as quasi-Newton methods to deal with the penalised problem (4.18), $\Gamma(X, \gamma_1, \gamma_2)$ needs to preserve the continuity of the second derivatives. Thus, the exterior penalty method, which preserves the continuity of any order derivatives of equality constraints, may be used to define the penalty function H_1, i.e.

$$H_1(X) = \sum_{j=1}^{m_1} [h_j(X)]^2 \tag{4.19}$$

To construct $H_2(X)$ for the inequality constraints, the quadratic extended interior penalty technique can be adopted, which is defined by

$$H_2(X) = \sum_{i=1}^{m_2} p_i(X) \tag{4.20}$$

where

$$p_i(X) = \begin{cases} -\dfrac{1}{g_i(X)} & \text{if } g_i(X) \leq -g_0 \\ \dfrac{1}{g_0}\left[\left[\dfrac{g_i(X)}{g_0}\right]^2 + 3\dfrac{g_i(X)}{g_0} + 3\right] & \text{if } g_i(X) \geq -g_0 \\ g_0 = c_2\sqrt{\gamma_2} & \end{cases} \tag{4.21}$$

g_0 is a positive transition parameter and c_2 is a positive constant.

In (4.20) and (4.21), the penalty function $H_2(X)$ is defined as an interior penalty function in most of the feasible design domain. It is defined as a quadratic exterior penalty function in a small part of the feasible domain (i.e. $-g_0 \leq g_i(X) \leq 0$) and in the infeasible domain ($g_i(X) \geq 0$). The penalty function $H_2(X)$ is continuous up to its second derivatives throughout the design space.

It can then be concluded that the pseudo-objective function $\Gamma(X, \gamma_1, \gamma_2)$ preserves the continuity of the first and the second derivatives of the functions in the original problem (4.6). It is therefore possible to use powerful unconstrained optimisation methods, which may require second order derivatives, to solve the penalised problem $P(\gamma_1, \gamma_2)$. The quasi-Newton methods are one class of methods which have superlinear convergence property and are well suited to dealing with the penalised problem.

The main principle behind the quasi-Newton methods consists of a generalisation of the recurrence formula of Newton's method

$$X^{k+1} = X^k - \lambda_k [\nabla^2 \Gamma(X^k)]^{-1} \nabla \Gamma(X^k) \tag{4.22}$$

where $\nabla^2 \Gamma(X^k)$ and $\nabla \Gamma(X^k)$ are the Hessian matrix and gradient vector of Γ at X^k and λ_k is the step size (a scalar).

The important limitation of Newton's method is the restriction that $\nabla^2 \Gamma(X^k)$ be positive definite. To address this restriction, the quasi-Newton methods consist of replacing $[\nabla^2 \Gamma(X^k)]^{-1}$ by a positive definite matrix H_k which is used to compute the direction of displacement from the gradient $\nabla \Gamma(X^k)$. The quasi-Newton methods therefore have a general iterative formula of the following type

$$X^{k+1} = X^k - \lambda_k H_k \nabla \Gamma(X^k) \tag{4.23}$$

where λ_k is chosen so as to minimise the function $\Gamma(X^k + \lambda_k d_k)$ in the direction $d_k = -H_k \nabla \Gamma(X^k)$.

The matrix H_k should obviously be modified at every step, so that H_k can be considered, at each step, to be a (positive definite) approximation to the inverse of the Hessian of Γ. The following general condition is then imposed on H_k

$$H_k^{-1}(\nabla \Gamma(X^k) - \nabla \Gamma(X^{k-1})) = X^k - X^{k-1} \tag{4.24}$$

Various correction formulae of the following type have been proposed

$$H_{k+1} = H_k + \Delta_k \tag{4.25}$$

where Δ_k is a matrix of rank 1 or of rank 2. Let

$$\delta_k = X^{k+1} - X^k \quad \text{and} \quad \pi_k = \nabla\Gamma(X^{k+1}) - \nabla\Gamma(X^k) \tag{4.26}$$

One of the well-known algorithms for updating H_k is the algorithm of Davidon-Fletcher-Powell (DFP). This algorithm uses the following correction formula

$$H_{k+1} = H_k + \frac{\delta_k \delta_k^T}{\delta_k^T \pi_k} - \frac{H_k \pi_k \pi_k^T H_k}{\pi_k^T H_k \pi_k} \tag{4.27}$$

The above algorithm can guarantee that H_{k+1} be positive definite. Suppose the matrix H_k is positive definite. If the following condition holds

$$\delta_k^T \pi_k > 0 \tag{4.28}$$

then the matrix H_{k+1} given by (4.27) is positive definite. Condition (4.28) holds if the point X^{k+1} is obtained from X^k by one-dimensional minimisation in the direction $d_k = -H_k \nabla\Gamma(X^k)$. The property of preserving the positive definiteness is essential because it ensures that the direction d_k, successfully generated by the algorithm, is a descent direction.

The DFP algorithm is rather sensitive to the accuracy in the one-dimensional optimisation. The algorithm of Broyden, Fletcher, Goldfarb and Shanno (BFGS) avoids this drawback. The correction formula of the BFGS algorithm is given by

$$H_{k+1} = H_k + \left[1 + \frac{\pi_k^T H_k \pi_k}{\delta_k^T \pi_k}\right] \frac{\delta_k \delta_k^T}{\delta_k^T \pi_k} - \frac{\delta_k \pi_k^T H_k + H_k \pi_k \delta_k^T}{\delta_k^T \pi_k} \tag{4.29}$$

The quasi-Newton methods with the DFP or BFGS correction formula can then be listed as follows.

Step 1: Select a starting point. Choose any positive definite H_0 (for example, the identity matrix) and let k=0.

Step 2: Calculate the displacement direction

$$d_k = -H_k \nabla\Gamma(X^k) \tag{4.30}$$

Find X^{k+1} as the minimum of

$$\Gamma(X^k + \theta d_k) \text{ for } \theta \geq 0 \tag{4.31}$$

Step 3: Calculate δ_k and π_k using formulae (4.26). Then, calculate H_{k+1} using either formula (4.27) (the DFP algorithm) or formula (4.29) (the BFGS algorithm).

Step 4: Let $k = k+1$. If X^{k+1} is not significantly different from X^k, stop the algorithm. Otherwise, go to step 2.

In the above algorithm, the positive definiteness of the matrices H_k is preserved if condition (4.28) holds. If Γ is a quadratic function with positive definite Hessian matrix A, then the formulae (4.27) and (4.29) make it possible to obtain, in at most n steps, the inverse A^{-1} of the Hessian of Γ. Furthermore, the directions δ_k successively generated by the algorithm are mutually conjugate with regard to A^{-1}.

However, the global convergence of the algorithm is only guaranteed if the algorithm is periodically restarted. For instance, one can take, after n steps, the last point obtained as the starting point and restart the matrix H (for example by setting $H_0 = I$, the identity matrix).

For given values of the penalty coefficients γ_1 and γ_2, $\Gamma(X, \gamma_1, \gamma_2)$ describes the bound of the feasible design space, referred to as the response surface. As the penalty coefficients γ_1 increases and γ_2 decreases, the contours of the response surface conform with the original objective function and the constraint functions more closely. It has been proven that the minima of the penalised problem (4.18) converge to the minima of the original problem (4.6) when $\gamma_1 \to \infty$ and $\gamma_2 \to 0$.

Problem (4.18) is solved for given γ_1 and γ_2. As γ_1 increases and γ_2 decreases, a series of the penalised problems are formulated and solved using for example the quasi-Newton methods until the minimum value of the pseudo-objective $\Gamma(X, \gamma_1, \gamma_2)$ coincides with the value of the original objective function $f(X)$ at $\overline{X}(\gamma_1, \gamma_2)$. It has been proven that $H_1(X) \to 0$ when $\gamma_1 \to \infty$ and $\gamma_2 H_2(X) \to 0$ when $\gamma_2 \to 0$. Therefore, $\overline{X}(\gamma_1, \gamma_2)$ may be regarded as a good approximation to the optimum of the original problem if both the value $H_1(\overline{X}(\gamma_1, \gamma_2))$ and the value $\gamma_2 H_2(\overline{X}(\gamma_1, \gamma_2))$ are sufficiently small. In numerical calculation, the convergence is normally tested by comparing the penalty functions with the original objective function. The following criteria are thus employed to test the convergence of the constrained optimisation

$$\varepsilon_{\gamma_1} = \left[\frac{H_1(X)}{f(X)}\right]^2 \leq \delta_{\gamma_1} \tag{4.32}$$

$$\varepsilon_{\gamma_2} = \left[\frac{\gamma_2 H_2(X)}{f(X)}\right]^2 \leq \delta_{\gamma_2} \tag{4.33}$$

where δ_{γ_1} and δ_{γ_2} are small positive numbers. Thus, $\varepsilon_{\gamma_1} \to 0$ when $\gamma_1 \to \infty$ and $\varepsilon_{\gamma_2} \to 0$ when $\gamma_2 \to 0$.

The penalty method for constrained nonlinear optimisation with the quasi-Newton methods can then be summarised as follows.

Step 1: Define a constrained optimisation problem as shown in (4.12).

Step 2: Assign initial values for the penalty coefficients γ_1^0 and γ_2^0, say, $\gamma_1^0 = \gamma_2^0 = 1$. Then, construct the penalised problem $P(\gamma_1^\rho, \gamma_2^\rho)$ as shown in (4.18) where $H_1(X)$ and $H_2(X)$ are defined by (4.19) and (4.20). Let $\rho = 0$.

Step 3: Implement the quasi-Newton methods to solve the $P(\gamma_1^\rho, \gamma_2^\rho)$. The minimum of the $P(\gamma_1^\rho, \gamma_2^\rho)$ is denoted by $\overline{X}^\rho = \overline{X}(\gamma_1^\rho, \gamma_2^\rho)$.

Step 4: If the convergence criteria (4.32) and (4.33) are both satisfied, $\overline{X}^\rho$ is regarded as a good approximation to the minimum $\overline{X}^*$ of (4.12). Then, let $\overline{X}^* = \overline{X}^\rho$ and stop.

Step 5: If (4.32) is not satisfied, let $\gamma_1^{\rho+1} = S_f \gamma_1^\rho$; if (4.33) is not satisfied, let $\gamma_2^{\rho+1} = \gamma_2^\rho / S_f$, where S_f is a scale factor larger than one, say $S_f = 10$. Let $\rho = \rho+1$ and $X^0 = \overline{X}^\rho$ and then go to Step 3.

4.3 Typical MODM Methods

4.3.1 Goal Programming

Goal programming (GP) was originally proposed by [Charnes and Cooper 1977] and has been developed further by other researchers. GP is a MODM technique which requires preference information before any efficient designs are generated. In fact, the method requires a designer to set goals for all objectives that he wishes to achieve. It adopts the decision rule that the best compromise design should be the one which minimises the weighted sum of deviations from the set goals. It allows the designer to assign preemptive weights to objectives and to define different achievement levels of the goals.

If the designer can provide goals for all objectives and accepts the above decision rule, GP may be one of the best methods to search for the best compromise solution. The computational steps of GP may be listed as follows.

Step 1: Define a MODM problem as in (4.1).

Step 2: Set goal values $\hat{f}_i$ for all objectives $f_i(X)$, $i=1, \cdots, k$.

Step 3: Assign preemptive weights p_l to objectives where $p_l \gg p_{l+1}$. This means that no number w, however large, can make $wp_{l+1} > p_l$. In other words, $f_i(X)$ will be regarded to be absolutely more important than $f_k(X)$ if $f_i(X)$ and $f_k(X)$ have preemptive weights p_l and p_{l+1}, respectively.

Step 4: Assign relative weights to objectives which have the same preemptive weight.

Step 5: Indicate whether each goal value should be attained as closely to as possible or is to be under-achieved or over-achieved.

Step 6: Use the above preference information to construct the GP problem for the MODM problem as follows

$$\min \quad \sum_{l=1}^{L} p_l \left[\sum_{j=j_1}^{j_l} (w_j^+ d_j^+ + w_j^- d_j^-) \right]$$

$$s.t. \quad X_b \in \Omega_b \tag{4.34}$$

$$\Omega_b = \left\{ X_b \left| \begin{array}{ll} f_j(X) - d_j^+ + d_j^- = \hat{f}_j & j=1, \cdots, k \\ d_j^+ . d_j^- = 0 & X_b = [X^T \; d_1^+ \; d_1^- \cdots d_k^+ \; d_k^-]^T \\ d_j^+ , \; d_j^- \geq 0 & X \in \Omega \end{array} \right. \right\}$$

where d_j^-, d_j^+ are deviation variables representing under-achievement and over-achievement of the goal $\hat{f}_j$. L is the number of the priority levels and j_l is the number of the objectives at a single priority level with the preemptive weight p_l.

Step 7: Problem (4.34) may be solved using the following sequence. Suppose $a_l(D^+, D^-)$ is the sum of deviations of the objectives at the lth priority level, defined by

$$a_l(D^+, D^-) = \sum_{j=j_1}^{j_l} (w_j^+ d_j^+ + w_j^- d_j^-) \tag{3-35}$$

where $D^+=[d_1^+ \cdots d_k^+]^T$ and $D^-=[d_1^- \cdots d_k^-]^T$. Let a_1^* be the optimal value of $a_1(D^+, D^-)$ obtained by solving the following problem

$$\begin{aligned} \min \quad & a_1(D^+, D^-) \\ s.t. \quad & X_b \in \Omega_b \end{aligned} \tag{4.36}$$

Then, solve the following $(L-1)$ problems sequentially using the SLP method or the penalty methods,

$$\begin{aligned} \min \quad & a_l \\ s.t. \quad & X_b \in \Omega_l \qquad l = 2, \cdots, L \end{aligned} \tag{4.37}$$

$$\Omega_l = \left\{ X_b \;\middle|\; \begin{array}{l} X_b \in \Omega_b \\ a_i(D^+, D^-) \le a_i^* \end{array} \quad i=1, \cdots, l-1 \right\}$$

If $X_b^* = [X^{*T} \ (d_1^+)^* \ (d_1^-)^*, \cdots, (d_k^+)^* \ (d_k^-)^*]^T$ is the optimal solution of problem (4.37) with $l=L$, then X^* is the best compromise solution of the MODM problem (4.1).

The goal programming is a widely used method. However, this method, like the simple weighting method, is not able to discover all efficient solutions if problem (4.1) is non-convex. Besides, in engineering design, it may not be easy for a designer to set goals for objectives especially when the objectives reflect technical performances of a new design problem.

4.3.2 Geoffrion's Method

Geoffrion's method is a search-oriented interactive method [Geoffrion et al. 1972]. It requires that the feasible decision space be convex and compact. Moreover, it assumes that a differentiable and concave utility function $u[f_1(X), \cdots, f_k(X)]$ exists but is not known explicitly. This method is an adoption of the Franke-Wolfe algorithm to multiobjective case. It only asks for such local information as is needed to perform the computations. The computational steps of the method may be summarised as follows.

Step 1 : Define a MODM problem as in (4.1) with "optimisation" being replaced by "maximisation".

Step 2: Choose an initial feasible or infeasible solution X^0 arbitrarily. Let $F^0 = F(X^0)$ where $F^0 = [f_1^0 \cdots f_i^0 \cdots f_k^0]^T$ and $f_i^0 = f_i(X^0)$ $(i=1, \cdots, k)$. Select an objective as the reference function denoted by f_r^0. Let t=0.

Step 3: Estimate the marginal rates of substitution (or indifferent trade-offs) w_{lr}^t between an objective f_l and the reference objective f_r at the solution X^t. w_{lr}^t is defined by

$$w_{lr}^t = \frac{\partial u(f^t(X))}{\partial f_l(X)} \Big/ \frac{\partial u(f^t(X))}{\partial f_r(X)} \Bigg|_{X=X^t} \tag{4.38}$$

The following procedure may be used to approximate w_{lr}^t through indifference tradeoff analysis. The designer is asked to compare the following two solutions

$$F^t = [f_1^t \quad \cdots \quad f_r^t \quad \cdots \quad f_l^t \quad \cdots \quad f_k^t]^T \tag{4.39}$$

and

$$\hat{F}^t = [f_1^t \quad \cdots \quad f_r^t + \Delta_r^t, \quad \cdots \quad f_l^t - \Delta_l^t \quad \cdots \quad f_k^t]^T \tag{4.40}$$

where Δ_r^t and Δ_l^t are small perturbations for f_r and f_l. If the designer prefers F^t to $\hat{F}^t$ (or $\hat{F}^t$ to F^t), Δ_l^t or Δ_r^t is regulated until indifference is reached. Then w_{lr}^t is given by Δ_r^t/Δ_l^t.

Step 4: Search for a direction along which the utility function u may be improved from X^t. First construct and solve the following direction problem.

$$\begin{aligned} \max \quad & \sum_{j=1}^{k} w_{jr}^t \nabla f_j(X^t) Y \\ s.t. \quad & Y \in \Omega \end{aligned} \tag{4.41}$$

Suppose Y^t is the optimal solution of (4.41). Then $d^t = Y^t - X^t$ is a direction along which the utility function can be improved.

Step 5: Formulate the following step-size problem

$$\begin{aligned} \max \quad & u[f_1(X^t+td^t), \cdots, u[f_k(X^t+td^t)] \\ s.t. \quad & 0 \le t \le 1 \end{aligned} \tag{4.42}$$

in order to determine the best step size at which the utility function is maximised along the direction d^t. Since u is not known explicitly, however, the solution of (4.42) can only be judged subjectively by the designer. One way for acquiring the judgement is to construct a table such as Table 4.1. Then the designer is required to select a t^q from the table at which the values of all the objectives are most preferred.

Table 4.1 *The Values of Criteria*

t^q	0	0.1	0.2	$\cdots$	0.9	1
$f_1(X^t+t^q d^t)$	f_1^0	$f_1^{0.1}$	$f_1^{0.2}$	$\cdots$	$f_1^{0.9}$	f_1^1
$f_2(X^t+t^q d^t)$	f_2^0	$f_2^{0.1}$	$f_2^{0.2}$	$\cdots$	$f_2^{0.9}$	f_2^1
$\cdots$	$\cdots$	$\cdots$	$\cdots$	$\cdots$	$\cdots$	$\cdots$
$f_k(X^t+t^q d^t)$	f_k^0	$f_k^{0.1}$	$f_k^{0.2}$	$\cdots$	$f_k^{0.9}$	f_k^1

Suppose t^* is the chosen step size, then $[f_1(X^t+t^*d^t), \cdots, f_k(X^t+t^*d^t)]^T$ is the vector of objective values of the best solution among those listed in Table 4.1.

Step 6: If $X^{t+1} = X^t$ or $F(X^{t+1}) = F(X^t)$, the iteration is stopped. However, such a theoretical convergency criterion is not easy to satisfy. So the following approximate criterion is generally applied

$$\frac{\Delta^t}{\Delta^0} = \frac{|(W^t)^T[F(X^{t+1}) - F(X^t)]|}{|(W^0)^T[F(X^1) - F(X^0)|} \le \alpha \tag{4.43}$$

where α is a threshold value preassigned by the designer. In (4.43), Δ^t/Δ^0 represents the ratio of the improvement of the utility function at interaction $t+1$ and that at interaction 1. If (4.43) is satisfied, the interaction is stopped. Otherwise, let $t=t+1$ and go to step 3.

As a search-oriented approach, Geoffrion's method is based on the strong assumption that the preferences provided by the designer must be consistent with an implicit utility function u. This assumption makes it difficult for a designer to provide preference information in a consistent manner. For instance, the determination of t^* becomes difficult when $k > 3$. In addition, the solution selected from Table 4.1, $X^{t+1} = X^t + t^*d^t$, may not be an efficient solution. Finally, the termination condition (4.43) is not a theoretical optimality condition. It may cause premature termination of an interactive decision making process.

4.3.3 Minimax Method

The minimax (or ideal point) method uses the distance between an efficient design and a pre-defined ideal design as the representation of a designer's overall preferences. It attempts to find from the feasible domain an efficient design which is nearest to the ideal design in the minimax sense.

The distance employed by the minimax method is defined as a weighted ∞-norm. A weighted p-norm is given by

$$d_p = \| \overline{W}(F(X) - F^-) \|_p = \left[\sum_{i=1}^{k} (\overline{w}_i (f_i(X) - f_i^-))^p \right]^{1/p} \quad (4.44)$$

where $F^- = [f_1^- \cdots f_i^- \cdots f_k^-]^T$, f_i^- is the minimum of $f_i(X)$ and $\overline{W} = [\overline{w}_1 \cdots \overline{w}_i \cdots \overline{w}_k]^T$ is the weighting and normalising vector with $\overline{w}_i$ being given by

$$\overline{w}_i = w_i / (f_i^* - f_i^-) \quad (4.45)$$

In (4.45), w_i is the relative weight of $f_i(X)$ and f_i^* is a value of $f_i(X)$ with $f_i^* > f_i^-$. Note that

$$\| \overline{W}(F(X) - F^-) \|_\infty = \max_i \{\overline{w}_i (f_i(X) - f_i^-)\} \quad (4.46)$$

The minimax problem is then defined by

$$\min_X \max_i \ \{\overline{w}_i (f_i(X) - f_i^-)\}$$

$$s.t. \quad X \in \Omega \quad (4.47)$$

The geometric significance of the minimax method is as illustrated by Figure 4.12 and 4.13. The contour of the ∞-norm in the objective space as given by equation (4.44) (p=∞) with regard to $F(X)$ is a set of hyperrectangles with F^- as the geometric centre. Solving problem (4.47) is to find the smallest hyperrectangle that just touches the feasible objective space Λ. If Λ is convex, it is always a vertex of the smallest hyperrectangle that just touches Λ as shown by point $F^1 = [f_1(X^1) \cdots f_k(X^1)]^T$ in Figure 4.12. In this case, the following equations hold

$$\overline{w}_1(f_1(X^1) - f_1^-) = \cdots = \overline{w}_i(f_i(X^1) - f_i^-) = \cdots = \overline{w}_k(f_k(X^1) - f_k^-) \quad (4.48)$$

In Figure 4.12, equation (4.48) is a line denoted by $\bar{w}_1(f_1(X^1) - f_1^-) = \bar{w}_2(f_2(X^1) - f_2^-)$, which passes through the two points F^- and F^1. If Λ is non-convex, it is possible that an edge of the smallest hyperrectangle may just touches Λ as shown by Figure 4.13. In the case of Figure 4.13, not all of equations (4.48) hold.

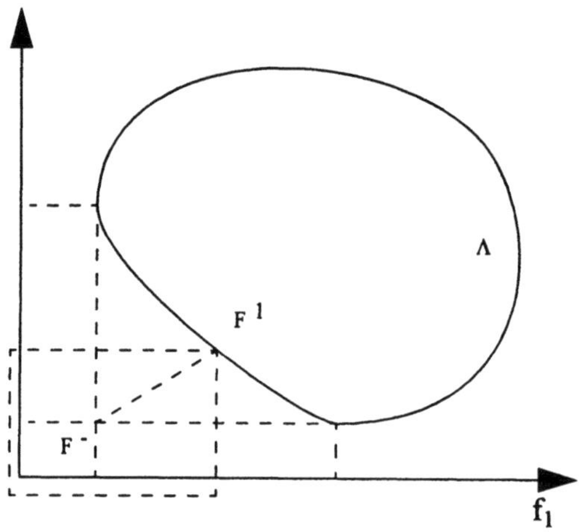

Figure 4.12 *Generation of Efficient Solutions by The Minimax Method (Convex Case)*

The computational steps of the minimax method are summarised as follows.

Step 1: Define a MODM problem as follows

$$\begin{aligned} \min \quad & F(X) = [f_1(X) \;\; f_2(X) \;\; \cdots \;\; f_k(X)]^T \\ s.t. \quad & X \in \Omega \end{aligned} \tag{4.49}$$

where Ω is defined in (4.1)

Step 2: Solve the following single-objective optimisation problems

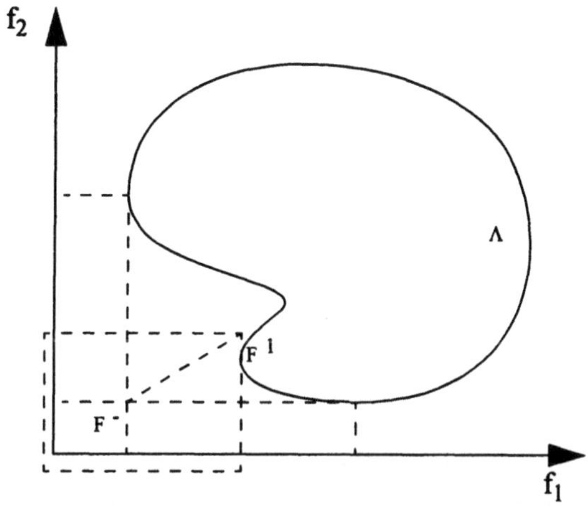

Figure 4.13 *Generation of Efficient Solutions by The Minimax Method (Non-Convex Case)*

$$\begin{aligned} \min \quad & f_i(X) \\ s.t. \quad & X \in \Omega \qquad i=1, \cdots, k \end{aligned} \tag{4.50}$$

Suppose the optimal solution of (4.50) is $\overline{X^i}$ and the values of the objectives at $\overline{X^i}$ are $f_j^- = f_j(\overline{X^i})$, $j=1, \cdots, k$. Then construct the pay-off table as follows.

Table 4.2 *Pay-off Table*

	$f_1(X)$	$f_2(X)$	$\cdots$	$f_k(X)$
$\overline{X^1}$	$f_1(\overline{X^1})$	$f_2(\overline{X^1})$	$\cdots$	$f_k(\overline{X^1})$
$\overline{X^2}$	$f_1(\overline{X^2})$	$f_2(\overline{X^2})$	$\cdots$	$f_k(\overline{X^2})$
$\cdots$	$\cdots$	$\cdots$		$\cdots$
$\overline{X^k}$	$f_1(\overline{X^k})$	$f_2(\overline{X^k})$	$\cdots$	$f_k(\overline{X^k})$

Step 3: Define the ideal point in the objective space as follows

$$F^{-} = [f_1(\overline{X}^1)\ f_2(\overline{X}^2)\ \cdots\ f_k(\overline{X}^k)]^T \tag{4.51}$$

Step 4: Use SLP or the quasi-Newton methods to solve the following problem, which is equivalent to problem (4.47), in order to generate an efficient solution

$$\begin{aligned} &\min\ \lambda \\ &s.t.\ \ \overline{w}_i(f_i(X) - f_i^{-}) \leq \lambda \qquad i = 1, 2, \cdots, k \\ &\qquad X \in \Omega,\ \ \lambda \geq 0 \end{aligned} \tag{4.52}$$

where f_i^* may be given from Table 4.2 as follows

$$f_i^* = \max_j\{f_i(\overline{X}^j)\} \qquad i = 1, 2, \cdots, k \tag{4.53}$$

Step 5: Suppose the optimal solution of problem (4.52) is X^t. If $\overline{W}$ represent the DM's overall preferences, X^t ought to be the best compromise solution. Otherwise, either w_i or f_i^* can be regulated to generated other efficient solutions. This leads to different interactive minimax procedures.

The minimax method is capable of discovering all efficient solutions of a multiobjective problem whether the problem is convex or non-convex. Given the relative weights of objectives, the minimax method can provide a compromise solution that is nearest the ideal solution. Obviously, the quality of the compromise solution obtained in this way depends on the accuracy of the weights. Based on such a solution, however, interactive procedures could be designed to conduct sensitivity analysis. This idea is adopted in the ISTM method as describeed below.

4.3.4 ISTM Method

The Interactive Step Trade-off Method (ISTM) is a learning-oriented interactive method [Yang et al. 1988, 1990]. It allows the DM to investigate the efficient frontier of a MODM problem by means of implicit trade-off analysis. In ISTM, objectives are classified into three sub-sets at a feasible solution. The DM is required to decide whether an objective has to be improved from the current level, or should be kept at least at the current level, or may be sacrificed from the current level. It is also necessary for him to indicate the extent to which an objective could be sacrificed from the current level. Based on the trade-off analysis, ISTM will try to find a new efficient design which can satisfy the designer's requirements and improve the assigned objectives to the largest possible extent.

The ISTM method provides a natural way of searching for good efficient solutions from which the best compromise design may be evolved. The computational steps of ISTM can be summarised as follows.

Step 1: Define a MODM problem as in (4.1) with "optimisation" being replaced by "maximisation".

Step 2: Use the minimax method to generate an efficient solution X^0 for given values of w_i's (i=1, $\cdots$, k). Let t = 1.

Step 3: Query the DM to classify the set of objective indices into the following three subsets,

W the index subset of objective functions which have to be improved from the current level $f_i(X^{t-1})$,

R the index subset of objective functions which should be kept at least at the current level $f_j(X^{t-1})$,

Z the index subset of objective functions which may be sacrificed from the current level $f_l(X^{t-1})$.

Let

$$\begin{cases} W = \{ i \mid i = i_1, i_2, \cdots, i_w \} \\ R = \{ j \mid j = j_1, j_2, \cdots, j_r \} \\ Z = \{ l \mid l = l_1, l_2, \cdots, l_z \} \end{cases} \tag{4.54}$$

where $W \cup R \cup Z = \{1, 2, \cdots, k\}$ and $W \cap R \cap Z = \varnothing$. Suppose $df_l(X^{t-1})$ is the current maximum decrement of $f_l(X)(l \in Z)$, and $df_l(X^{t-1}) \geq 0$.

Step 4: Suppose $u_i (i \in W)$ is an auxiliary variable, then an auxiliary problem can be defined as follows

$$\begin{aligned} \max \quad & u = \sum_{i \in W} \sigma_i u_i \\ s.t. \quad & X_a \in \Omega_a \end{aligned} \tag{4.55}$$

$$\Omega_a = \left\{ X_a \left| \begin{array}{ll} f_i(X) - h_i u_i \geq f_i(X^{t-1}),\ u_i \geq 0 & i \in W \\ f_j(X) \geq f_j(X^{t-1}) & j \in R \\ f_l(X) \geq f_l(X^{t-1}) - df_l(X^{t-1}) & l \in Z \\ X_a = [X^T \ u_{i_1} \ \cdots \ u_{i_w}]^T & X \in \Omega \end{array} \right. \right\}$$

where u_i is maximised to improve $f_i(X)$ as greatly as possible, u the auxiliary objective function, and σ_i a positive weighting factor which is determined according to the relative importance of the objective functions in the subset W. Normally, we let $\sigma_i=1$ $(i \in W)$. h_i is a weighting factor which can be given by

$$h_i = |f_i^* - f_i^-| \qquad i \in W \tag{4.56}$$

where f_i^* is given by the optimal solution of the following problem, denoted by $\overline{X}^i$

$$\begin{aligned} &\max \ f_i(X) \\ &s.t. \ X \in \Omega \end{aligned} \tag{4.57}$$

or $f_i^* = f_i(\overline{X}^i)$ for all $i=1, \cdots, k$. f_i^- is given by

$$f_i^- = \min_j \{f_i(\overline{X}^j)\} \qquad i=1, \cdots, k \tag{4.58}$$

Step 5: Use the SLP method or one of the quasi-Newton methods to solve the auxiliary problem (4.55). The optimal solution of (4.55), denoted as X^t, is a new (weakly) efficient solution of the original MODM problem [Yang et al. 1990]. Then the DM is required to evaluate X^t.

Step 6: If the designer is unsatisfied with X^t, let $t = t + 1$ and then go to step 3. This interactive process is terminated if a> no objective is required to improve, or b> no objective is allowed to sacrifice.

Figure 4.14 shows graphically how ISTM works and the interpretation of the implicit trade-offs. In Figure 4.14, the feasible solution space of a problem with two objectives is the area enclosed by line $\overline{XO}$, curve $\tilde{XY}$ and line $\overline{YO}$. The search space of the auxiliary problem (4.55) defined at X^{t-1} (point A) is the shaded area within the original feasible space. If objective 1 needs to be improved from its value at point A, this can only be done at the expense of objective 2. If S is the limiting value of sacrifice for objective 2, the new solution is B.

This trade-off also implies a change in relative importance of the two objectives denoted by a shift from line aa to line bb, which are the tangent lines of the curve $\tilde{XY}$ at points A and B. It should be noted that the new efficient solution denoted by point B may never be discovered by some methods which can only deal with convex MODM problems such as the simple additive weighting method, goal programming or Geoffrion's method. However, ISTM is capable of

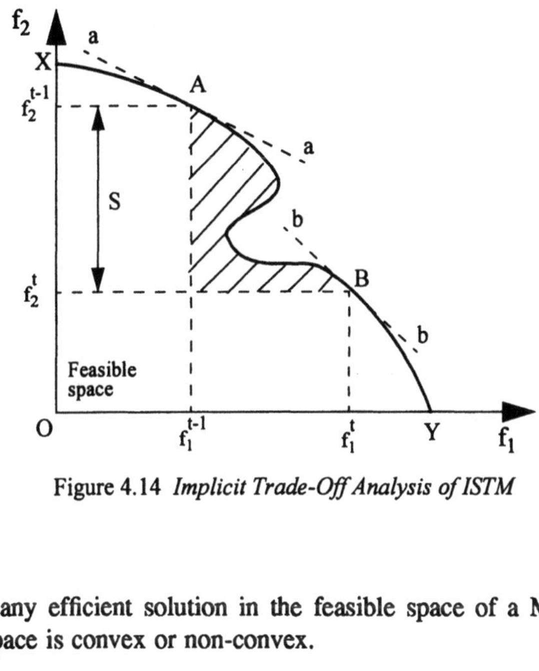

Figure 4.14 *Implicit Trade-Off Analysis of ISTM*

searching for any efficient solution in the feasible space of a MODM problem, whether the space is convex or non-convex.

Direct search of the efficient frontier for a MODM problem may be one of the most realistic ways of dealing with multiobjective preliminary design problems as there is often not enough *a priori* preference information available for a new design problem. Even an experienced designer may like to see as many feasible design options as possible before he decides which design he really prefers.

4.3.5 Local Utility Function Method

In the above ISTM method, the DM is expected to subjectively terminate an interactive process when he recognises that his best compromise solution has evolved already. Such a termination strategy may be implemented when the DM no longer wishes to improve an objective at the expense of any other objectives.

In this section, a search process is introduced so that an interactive process can be terminated in a more objective way. This search process is composed of the estimation and optimisation of a set of utility functions on the basis of preferences provided by the DM in terms of the pairwise comparisons between objectives and between efficient solutions [Yang and Sen 1996a].

The computational steps of this method can be summarised as follows.

Step 1: Define a MODM problem as in (4.1) with "optimisation" being replaced by "maximisation".

Step 2: Optimise each of the objectives to obtain the best value of each objective, denoted by f_i^* for objective $f_i(X)$. Suppose f_i^0 is the acceptable value of objective $f_i(X)$. Then, define $[f_i^0 \; f_i^*]$ as the acceptable interval of $f_i(X)$, and

$$\Omega_a = \{ X \mid f_i(X) \in [f_i^0 \; f_i^*], i=1, \cdots, k; \; X \in \Omega \} \tag{4.59}$$

as the acceptable decision space of Ω, or a local region of Ω ($\Omega_a \subseteq \Omega$). $X \in \Omega_a$ is referred to as an acceptable solution.

Step 3: Define an additive piecewise linear local utility function as follows

$$u(X) = \sum_{i=1}^{k} \left\{ \hat{u}_i(f_i^{j-1}) + \frac{f_i(X) - f_i^{j-1}}{f_i^j - f_i^{j-1}} \left[\hat{u}_i(f_i^j) - \hat{u}_i(f_i^{j-1}) \right], \right. \tag{4.60}$$

$$\text{if } f_i(X) \in [f_i^{j-1} \; f_i^j] \}$$

where $\hat{u}_i(f_i^j)$, a parameter to be estimated, is a marginal utility of f_i^j which is an end point of an equal sub-interval $[f_i^{j-1} \; f_i^j]$ as given by

$$f_i^j = f_i^0 + \frac{j}{N_i}(f_i^* - f_i^0) \quad j = 0, 1, \cdots, N_i \tag{4.61}$$

where $f_i^{N_i} = f_i^*$. Let f_i^- be the worst value of $f_i(X)$ with the lowest normalised utility of zero and $f_i^- \leq f_i^0$. N_i is the number of equal intervals for objective $f_i(X)$ and may be assigned so that $1 \leq N_i \leq 10$.

Step 4: Use the ISTM method or some other interactive method to generate a subset of acceptable solutions, as shown in Table 4.3, which are mutually comparable. Let Γ be the set of the generated acceptable solutions, or $\Gamma = \{X^0 \cdots X^h \cdots X^l \cdots X^T\}$. Let $\mathbf{P}$ be the strict preference relation and $\mathbf{I}$ the indifference relation between two solutions. The relation that X^l is preferred to X^h is then denoted by $X^l \mathbf{P} X^h$, and that X^l is indifferent to X^h by $X^l \mathbf{I} X^h$. Let Ω_P denote the set of all pairs of solutions with the preference relation $\mathbf{P}$ and Ω_I the set of all pairs of solutions with the preference relation $\mathbf{I}$, that is

$$\Omega_P = \{(X^l, X^h) \mid X^l \mathbf{P} X^h \text{ for } all\ X^l, X^h \in \Gamma\} \tag{4.62}$$

$$\Omega_I = \{(X^l, X^h) \mid X^l \mathbf{I} X^h \text{ for } all\ X^l, X^h \in \Gamma\} \tag{4.63}$$

Table 4.3 *Trade–Off Table* for *Preference Learning*

Efficient solutions Γ	Objectives				
	$f_1(X^t)$	$\cdots$	$f_i(X^t)$	$\cdots$	$f_k(X^t)$
X^0	$f_1(X^0)$	$\cdots$	$f_i(X^0)$	$\cdots$	$f_k(X^0)$
$\cdots$	$\cdots$	$\cdots$	$\cdots$	$\cdots$	$\cdots$
X^h	$f_1(X^h)$	$\cdots$	$f_i(X^h)$	$\cdots$	$f_k(X^h)$
$\cdots$	$\cdots$	$\cdots$	$\cdots$	$\cdots$	$\cdots$
X^l	$f_1(X^l)$	$\cdots$	$f_i(X^l)$	$\cdots$	$f_k(X^l)$
$\cdots$	$\cdots$	$\cdots$	$\cdots$	$\cdots$	$\cdots$
X^T	$f_1(X^T)$	$\cdots$	$f_i(X^T)$	$\cdots$	$f_k(X^T)$

Step 5: Estimate marginal utilities at the end points, i.e. $u_i(f_i^j)$ (j=0, 1, $\cdots$, N_i; i=1, $\cdots$, k), using the following linear goal programming model

$$\min\ P_1\left\{\sum_{(X^l, X^h)\in\Omega_P} d_{lh}^+ + \sum_{(X^l, X^h)\in\Omega_I} (d_{lh}^+ + d_{lh}^-) + \sum_{(i,j)\in\Omega_I} (s_{ij}^+ + s_{ij}^-)\right\}$$

$$+ P_2 \sum_{X^l\in\Gamma} (\sigma_l^+ + \sigma_l^-) \tag{4.64a}$$

$$s.t.\quad \sum_{i=1}^{k}\left\{\overline{u}_i(f_i(X^l)) - \overline{u}_i(f_i(X^h))\right\} + \sigma_l^+ - \sigma_l^- - \sigma_h^+ + \sigma_h^- + d_{lh}^+ \geq \delta$$

$$\text{if } (X^l, X^h) \in \Omega_P \tag{4.64b}$$

$$\sum_{i=1}^{k}\left\{\overline{u_i}(f_i(X^l)) - \overline{u_i}(f_i(X^h))\right\}+\sigma_l^+-\sigma_l^--\sigma_h^++\sigma_h^-+d_{lh}^+-d_{lh}^- = 0$$

$$\text{if } (X^l, X^h) \in \Omega_I \tag{4.64c}$$

$$u_i(f_i^j) \geq u_i(f_i^{j-1}) \geq 0; \ \text{ for } j=1, \cdots, N_i; \ i=1, \cdots, k \tag{4.64d}$$

$$\sum_{i=1}^{k} u_i(f_i^{N_i}) = 1;$$

$$u_i(f_i^0) - \frac{f_i^0 - f_i^-}{f_i^1 - f_i^0}\left[u_i(f_i^1) - u_i(f_i^0)\right] = 0 \ \text{ for } i=1, \cdots, k \tag{4.64e}$$

$$u_i(f_i^{N_i}) - c_{ij}u_j(f_j^{N_j}) + s_{ij}^+ \geq 0 \quad or \tag{4.64f}$$

$$u_i(f_i^{N_i}) - c_{ij}u_j(f_j^{N_j}) - s_{ij}^- + s_{ij}^+ = 0 \ \text{ for } all \ (i, j) \in \Omega_f \tag{4.64g}$$

$$\sigma_l^+ \times \sigma_l^- = 0; \ \ d_{lh}^+ \times d_{lh}^- = 0; \ \ s_{ij}^+ \times s_{ij}^- = 0 \tag{4.64h}$$

$$\sigma_l^+, \sigma_l^- \geq 0; \ \ d_{lh}^+, d_{lh}^- \geq 0; \ \ s_{ij}^+, s_{ij}^- \geq 0 \tag{4.64i}$$

$$\text{for } all \ X^l \in \Gamma; \ (X^l, X^h) \in \Omega_P \cup \Omega_I; \text{ and } (i, j) \in \Omega_f$$

where P_1 and P_2 represent priority weights for inconsistency errors and approximation errors, respectively, with $P_1 \gg P_2$. In problem (4.64), δ is a small positive constant, σ_l^+, σ_l^- for all $X^l \in \Gamma$ are non-negative approximation error variables, and d_{lh}^+, d_{lh}^- for all $(X^l, X^h) \in \Omega_P \cup \Omega_I$ and s_{ij}^+ and s_{ij}^- for all $(i, j) \in \Omega_f$ are non-negative inconsistency error variables where Ω_f is the index set of pairs of objectives with the importance of one objective comparable with that of the other, or more precisely $\Omega_f = \{(i, j) \mid f_i(X) \textit{ is comparable with } f_j(X) \ i, \ j=1, \cdots, k; \ i \neq j\}$. Note that $u_i(f_i^{N_i})$ is equivalent to the weight of $f_i(X)$. c_{ij} in equation (4.64f) or (4.64g) means that objective i is at least or exactly c_{ij} times as important as objective j. $\overline{u_i}(f_i(X^l))$ is the marginal utility of design X^l on the ith objective approximated by the following interpolation

$$\overline{u}_i(f_i(X^l)) = u_i(f_i^{j-1}) + \frac{f_i(X^l) - f_i^{j-1}}{f_i^j - f_i^{j-1}} \left[u_i(f_i^j) - u_i(f_i^{j-1}) \right]$$

$$\text{if } f_i(X^l) \in [f_i^{j-1} \ f_i^j] \tag{4.65}$$

The optimal solution of problem (4.64) is denoted by $\hat{u}_i(f_i^j)$, $\hat{d}_{lh}^+$, $\hat{d}_{lh}^-$, $\hat{\sigma}_l^+$, $\hat{\sigma}_l^-$, $\hat{s}_{ij}^+$ and $\hat{s}_{ij}^-$.

Step 6: If $\hat{d}_{lh}^+=\hat{d}_{lh}^-=0$ for all $(X^l, X^h)\in\Omega_P\cup\Omega_I$, $\hat{\sigma}_l^+=\hat{\sigma}_l^-=0$ for all $X^l\in\Gamma$ and $\hat{s}_{ij}^+=\hat{s}_{ij}^-=0$ for all $(i, j)\in\Omega_f$, the assessed optimal utility function (4.60) can precisely and consistently model the DM's preferences. In this case, there often exist other optimal utility functions which all lead to a perfect representation of the preferences. The following problems are therefore designed to identify the upper and lower bounds of the permissible optimal values for each marginal utility,

$$LP(\overline{u}_i^j) \begin{cases} \max & u_i(f_i^j) \\ s.t. & equations\ (4.64b) - (4.64g) \\ & with\ all\ error\ variables\ being\ zero \end{cases} \tag{4.66}$$

$$LP(\underline{u}_i^j) \begin{cases} \min & u_i(f_i^j) \\ s.t. & equations\ (4.64b) - (4.64g) \\ & with\ all\ error\ variables\ being\ zero \end{cases} \tag{4.67}$$

for j=0, 1, $\cdots$, N_i; i=1, $\cdots$, k. Let $N = \sum_{i=1}^{k}(N_i + 1)$. Solving problems (4.66) and (4.67), we can obtain $2N$ optimal solutions, the mean of which is also an optimal solution. The mean marginal utility functions are often smoother than individual optimal utility functions and may therefore be used in equation (4.60).

Step 7: The estimated optimal (or mean) utility functions may then be optimised within the acceptable region to search for the best compromise design. The following ordinary nonlinear programming problem is constructed for utility optimisation

$$\begin{aligned} \max\ & u(F(X)) = \sum_{i=1}^{k} \left\{ \sum_{j=1}^{N_i-1} \alpha_{ij}(a_{ij}^+ + a_{ij}^-) + \beta_i f_i(X) + \gamma_i \right\} \\ s.t.\ & X \in \Omega_a \\ & f_i(X) - a_{ij}^+ + a_{ij}^- = f_i^j \quad j=1, \cdots, N_i-1;\ i=1, \cdots, k \\ & a_{ij}^+ \times a_{ij}^- = 0;\ a_{ij}^+, a_{ij}^- \geq 0 \quad j=1, \cdots, N_i-1;\ i=1, \cdots, k \end{aligned} \tag{4.68}$$

where the utility function denoted by equation (4.60) can be represented by introducing auxiliary variables a_{ij}^{+} and a_{ij}^{-} and following coefficients

$$\alpha_{ij} = \frac{1}{2}(t_{i,j+1} - t_{i,j}),\ \beta_i = \frac{1}{2}(t_{i,1} + t_{i,N_i})$$

$$\text{and } \gamma_i = \frac{1}{2}(s_{i,1} + s_{i,N_i}) \tag{4.69}$$

$$t_{i,j} = \frac{\hat{u}_i(f_i^j) - \hat{u}_i(f_i^{j-1})}{f_i^j - f_i^{j-1}};\quad s_{i,1} = \hat{u}_i(f_i^0) - t_{i,1} f_i^0,$$

$$\text{and } s_{i,N_i} = \hat{u}_i(f_i^{N_i}) - t_{i,N_i} f_i^{N_i} \tag{4.70}$$

It is easy to prove that the optimal solution of problem (4.68) is an efficient solution of problem (4.1) if each estimated marginal utility function, say $u_i(f_i(X))$, is a strictly monotonically increasing function of $f_i(X)$. The best compromise solution obtained by solving problem (4.68) may only be preferred in a local sense in that the utility functions are obtained on the basis of the ordinal pairwise comparisons of the efficient solutions already generated. If the DM is not confident about or satisfied with this solution, he could undertake post-optimality analysis as follows, with a view to identifying a more satisfactory solution.

Suppose the pairwise comparisons of the generated efficient solutions are not consistent with one another so that some of the d_{lh}^{+} or d_{lh}^{-} variables are not zero. Then, the DM can modify those comparisons with non-zero values of d_{lh}^{+} or d_{lh}^{-} obtained. If the assumed piecewise linear utility function cannot precisely represent the DM's preferences so that some σ_l^{+} or σ_l^{-} in the solution of problem (4.64) is not zero, then the acceptable interval of an objective function needs to be divided into smaller equal sub-intervals. On the other hand, if the DM is not satisfied with the assignment of the relative weights of the objectives, equivalent to the estimated marginal utilities (i.e. $\omega_i = \hat{u}_i(f_i^*) = \hat{u}_i(f_i^{N_i})$ $(i=1, \cdots, k)$), he can assign a new set of weights using other methods for weight assignment. In fact, different sets of weights may be used to estimate different utility functions which may then be optimised to search for different compromise solutions, as discussed in the following sections.

4.4 Multiobjective Ship Design

4.4.1 A Nonlinear Preliminary Ship Design Model

The detailed design of a ship is a complex process. At early design stages, however, a mathematical ship design model is often useful for preliminary design synthesis and analysis. To investigate a design model using analytical tools together with general nonlinear optimisation algorithms, the mathematical functions of the model need to be continuous and differentiable. In engineering design, however, tables and diagrams are often used to represent some of the discrete relations between design variables and regression analysis may be used to arrive at approximate continuous functions.

The ship powering model used in this section deals with the preliminary design of bulk carriers. However the model is highly approximate and used in part outside its technical range so that the results are to be treated as comparative only and not as strictly correct from the naval architectural point of view. This approximate model works on the basis that for a given hull form and ship speed (and hence *Froude Number*) the power required to propel the vessel can be computed using a coefficient referred to as the *Admiralty Coefficient* [Yang and Sen 1996b]. For given values of a ship design variable referred to as *block coefficient* which represents fullness of form and is denoted by C_B, the relations between two parameters, *Admiralty Coefficient* (ACo) and *Froude Number* (FN), are as shown by Figure 4.15, where $ACo = \Delta^{2/3}V^3/P$, $FN = V/(gL)^{1/2}$ and Δ, V, L, P and g are ship displacement, speed, length, power and acceleration due to gravity, respectively. It is obvious from Figure 4.15 that ACo is a linear function of FN given C_B. Let $ACo = b(C_B)FN + a(C_B)$ where $b(C_B)$ and $a(C_B)$ are slope and y-intercept of a line given C_B. For each of the lines, two points are sampled as shown by Table 4.4 from which a and b can be calculated. The values of a and b for different values of C_B are shown in Table 4.5.

Figure 4.16 illustrate the variation of a and b with C_B. From Figure 4.16, it may be assumed that both $b(C_B)$ and $a(C_B)$ are quadratic functions of C_B of the following form

$$b(C_B) = \zeta_1 C_B^2 + \zeta_2 C_B + \zeta_3 \tag{4.71}$$

$$a(C_B) = \eta_1 C_B^2 + \eta_2 C_B + \eta_3 \tag{4.72}$$

Using a linear least square algorithm, the coefficients of equations (4.71) and (4.72) are estimated as

$$[\zeta_1 \quad \zeta_2 \quad \zeta_3]^T = [-10847.2 \quad 12817 \quad -6960.32]^T \tag{4.73}$$

$$[\eta_1 \quad \eta_2 \quad \eta_3]^T = [4977.06 \quad -8105.61 \quad 4456.51]^T \tag{4.74}$$

The maximum relative estimation error for each of the sampled data as shown in Table 3 is less than 1%.

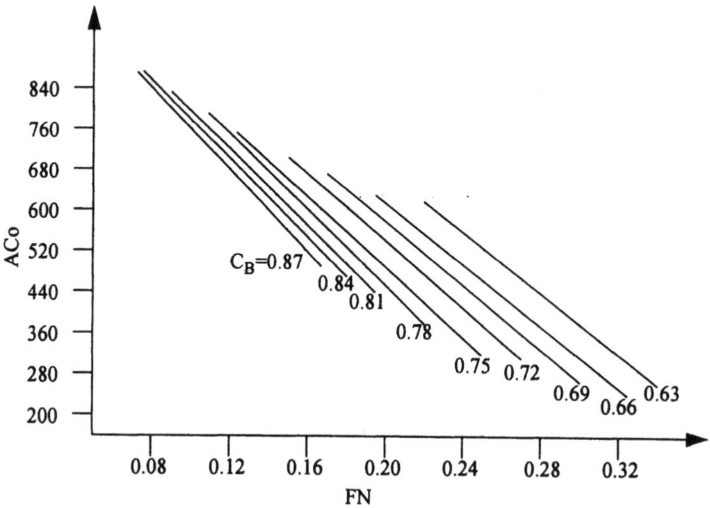

Figure 4.15 *Relation between FN* and AC_o

Table 4.4 *Sampled Data* for *ACo* and *FN*

C_B	point 1	point 2	C_B	point 1	point 2
0.87	(0.160, 520)	(0.120, 680)	0.72	(0.250, 360)	(0.190, 560)
0.84	(0.178, 480)	(0.100, 780)	0.69	(0.260, 372)	(0.190, 600)
0.81	(0.195, 440)	(0.120, 720)	0.66	(0.320, 236)	(0.220, 560)
0.78	(0.210, 420)	(0.115, 760)	0.63	(0.330, 280)	(0.230, 600)
0.75	(0.214, 440)	(0.150, 660)			

A ship design model is built to investigate a family of bulk carriers, ranging from 3,000 to 500,000 tonnes deadweight, with speeds ranging from about 14 to 18 knots. The model accommodates a range of technical requirements defining a design space and a few conflicting technical and economic performance indices for evaluation of each ship design.

Table 4.5 *The Changes of a and b with C_B*

C_B	b	a	C_B	b	a
0.87	-4000.00	1160.00	0.72	-3333.33	1193.33
0.84	-3846.15	1164.61	0.69	-3257.14	1218.86
0.81	-3733.33	1168.00	0.66	-3240.00	1272.80
0.78	-3578.95	1171.58	0.63	-3200.00	1336.00
0.75	-3437.50	1175.63			

Three performance indices (or objectives) are taken into consideration which are *tranportation cost*, *light ship mass* and *annual cargo carrying capacity* (simplely *annual cargo*). A ship design is favourable if it has low *tranportation cost*, low *light ship mass* and high *annual cargo*. Six independent design variables are defined which are *length* (L), *draft* (T), *depth* (D), *block coefficient* (C_B), *breadth* (B) and *speed* (V) where L, T, D and B are measured in meters, V in knots and C_B is a dimensionless variable. The three objectives are defined as follows [Yang and Sen 1996b].

- Minimisation of *tranportation cost* (TC)

$$tranportation\ cost = \frac{annual\ costs}{annual\ cargo} \tag{4.75}$$

- Minimisation of *light ship mass* (LSM)

$$light\ ship\ mass = steel\ mass + outfit\ mass + machinery\ mass \tag{4.76}$$

- Maximisation of *annual cargo* (AC)

$$annual\ cargo = cargo\ deadweight \times RTPA \tag{4.77}$$

The terms in the objectives are defined with regard to the six design variables as follows

$$annual\ costs = capital\ charges + running\ costs + voyage\ costs \times RTPA$$

$$capital\ charges = 0.2 \times ship\ cost$$

$$ship\ cost = 1.3 \times (2000 \times (steel\ mass)^{0.85} + 3500 \times (outfit\ mass) + 2400 \times P^{0.8})$$

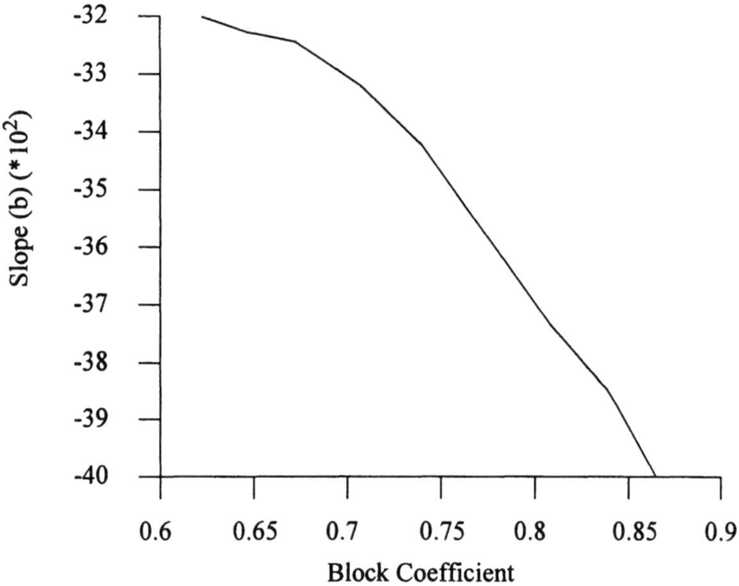

Figure 4.16a *Relation between* C_B and *b*

$$steel\ mass = 0.034 \times L^{1.7} \times B^{0.7} \times D^{0.4} \times C_B^{0.5}$$

$$outfit\ mass = 1.0 \times L^{0.8} \times B^{0.6} \times D^{0.3} \times C_B^{0.1}$$

$$machinery\ mass = 0.17 \times P^{0.9}$$

$$P = displacement^{2/3} \times V^3 \times \frac{1}{b(C_B) \times \dfrac{V}{(g \times L)^{0.5}} + a(C_B)}$$

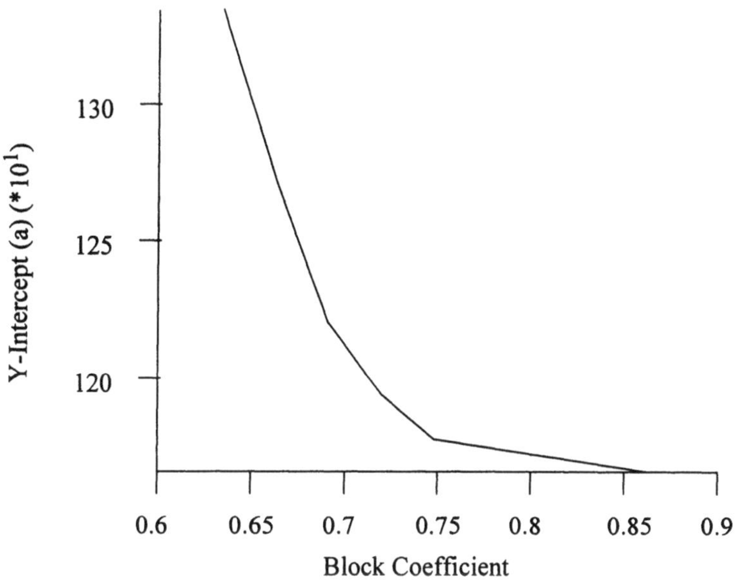

Figure 4.16b *Relation between* C_B and *a*

$$displacement = 1.025 \times L \times B \times T \times C_B$$

$$running\ costs = 40000 \times DW^{0.3}$$

$$DW = displacement - light\ ship\ mass$$

$$voyage\ costs = fuel\ cost + port\ cost$$

$$fuel\ cost = 1.05 \times (daily\ consumption) \times (sea\ days) \times (fuel\ price)$$

$$daily\ consumption = P \times 0.19 \times 24/1000 + 0.2$$

$$sea\ days = \frac{round\ trip\ miles}{24 \times V}$$

$$round\ trip\ miles = 5000 \qquad (nautical\ miles)$$

$$fuel\ price = 100 \qquad (pounds/tonne)$$

$$port\ cost = 6.3 \times DW^{0.8}$$

$$RTPA = \frac{350}{sea\ days + port\ days}$$

$$port\ days = 2 \times \left[\frac{cargo\ deadweight}{cargo\ handling\ rate} + 0.5 \right]$$

$$cargo\ deadweight = DW - fuel\ carried$$

$$- crew,\ stores\ \text{and}\ water$$

$$fuel\ carried = (daily\ comsumption) \times (sea\ days + 5)$$

$$crew,\ stores\ \text{and}\ water = 2.0 \times DW^{0.5}$$

$$cargo\ handling\ rate = 8000 \qquad (tonnes/day)$$

where $RTPA$ is *round trip per annum*, DW is *deadweight*, g is 9.8065 and $a(C_B)$ and $b(C_B)$ are given by equations (4.71) to (4.74).

A feasible design needs to satisfy the following technical requirements which are interpreted as the constraints on ship dimensions, displacement, powering and stability.

A. Dimensions and displacement

- Length/Breadth ratio

$$L/B \geq 6 \qquad (4.78)$$

- Length/Depth ratio

$$L/D \leq 15 \qquad (4.79)$$

- Length/Draft ratio

$$L/T \leq 19 \qquad (4.80)$$

- Draft constraints

$$T \leq 0.45 \times DW^{0.31} \ \text{and} \ T \leq 0.7 \times D + 0.7 \qquad (4.81)$$

- Deadweight constraint

$$3{,}000 \leq DW \leq 500{,}000 \tag{4.82}$$

B. Powering

- Block coefficient constraints

$$0.63 \leq C_B \leq 0.75 \tag{4.83}$$

- Speed constraints

$$14 \leq V \leq 18 \tag{4.84}$$

- Froude number constraint

$$V/(g \times L)^{0.5} \leq 0.32 \tag{4.85}$$

C. Stability

- Metacentric height (GM)

$$GM \geq 0.07 \times B \tag{4.86}$$

$$GM = KB + BM - KG$$

$$KB = 0.53 \times T$$

$$BM = \frac{(0.085 \times C_B - 0.002) \times B^2}{T \times C_B}$$

$$KG = 1.0 + 0.52 \times D$$

4.4.2 Generation of Subsets of Efficient Ship Designs

The ship design model as given above is a nonlinear multiobjective problem. The general assumptions are that (i) there is unlimited cargo availability and (ii) there is no capital rationing. The model can be generally formulated as a nonlinear vector optimisation problem. A design is denoted by $X = [x_1 \ x_2 \ x_3 \ x_4 \ x_5 \ x_6]^T = [L \ T \ D \ C_B \ B \ V]^T$ and the three normalised objectives *tranportation cost* (TC), *light ship mass* (LSM)/10,000 and *annual cargo* (AC)/1,000,000 by $f_1(X)$, $f_2(X)$ and $f_3(X)$, respectively, or

$F(X) = [-f_1(X)\ \ -f_2(X)\ \ f_3(X)]^T$. The model may then be generalised by

$$\begin{cases} \min & \{f_1(X), f_2(X)\} \\ \max & \{f_3(X)\} \\ s.t. & X \in \Omega \end{cases} \tag{4.87}$$

$$\Omega = \{X \mid \textit{any } X \textit{ which satisfies constraints } (4.78) \textit{ to } (4.86)\} \tag{4.88}$$

Each of the three objectives are optimised individually at first to generate the three extreme efficient designs as shown by Table 4.6 and denoted by $\overline{X}^1$, $\overline{X}^2$ and $\overline{X}^3$. The hull converges to the finest from permissible which is unusual for bulk carrier, but this is largely due to the high fuel price assumed in the model. It is obvious from Table 4.6 that the three extreme designs are significantly different from one another, which indicates that the three objectives are in some conflict. To demonstrate in more detail the mutual conflict between each pair of objectives, three subsets of the efficient designs can be sampled using the ISTM method, as shown by Figure 4.17.

Table 4.6 *Extreme Efficient Ship Designs*

extreme	variables						objectives		
designs	x_1	x_2	x_3	x_4	x_5	x_6	$f_1(X)$	$f_2(X)$	$f_3(X)$
$\overline{X}^1$	271.09	16.08	21.97	0.63	45.18	14.00	9.458	2.236	0.889
$\overline{X}^2$	195.15	10.27	13.67	0.63	24.14	14.00	12.814	0.716	0.372
$\overline{X}^3$	499.63	26.30	36.57	0.63	70.27	18.00	17.341	9.615	1.270

Figure 4.17a shows that the conflict between *LSM* and *TC* is relatively low as the minimisation of the latter only leads to 2.2355×10000 tonnes of *LSM* which is far below the heaviest *LSM* of 9.6145×10000 tonnes as given in Table 4.6 while the minimisation of *LSM* only leads to 12.814 pounds/tonne of *TC* well below the worst *TC* of 17.3413 pounds/tonne. In the same way, one can find that the conflict between *AC* and *TC* or *LSM* is high as shown by Figure 4.17b and Figure 4.17c.

4.4.3 Progressive Design

Due to the existence of conflict among the three objectives, no single design could optimise all the objectives simultaneously. Compromise is therefore necessary to find the best design with the three objectives being each given as

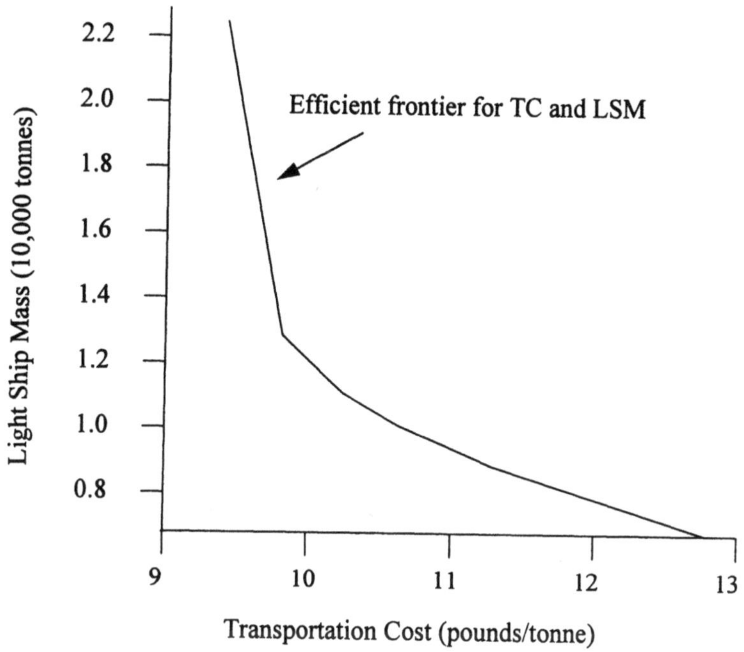

Figure 4.17a *Local Efficient Frontier between TC* and *LSM*

much importance as possible. Interactive trade-off analyses can be undertaken to arrive at a good design progressively using Geoffrion's method or a representative subset of efficient solutions can be generated using the ISTM method.

Following the procedure of Geoffrion's method as described in section 4.3.2, a compromise ship design can be approached in a progressive manner. The threshold value α in the termination criterion, as defined by equation (4.43), is given by 0.01, or $\alpha = 0.01$.

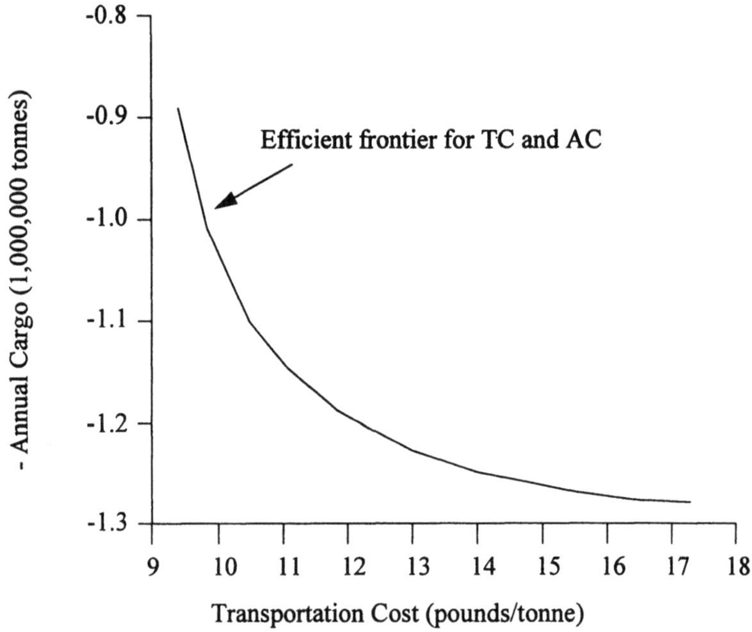

Figure 4.17b *Local Efficient Frontier between TC* and *AC*

Suppose an initial design is given arbitrarily, for example

$$X^0 = [200 \quad 17 \quad 25 \quad 0.7 \quad 30 \quad 15]^T$$
$$F(X^0] = [-19.1902 \quad -1.3725 \quad 0.6658]^T$$

Note that $F(X)$ is defined by

$$F(X) = [-f_1(X) \quad -f_2(X) \quad f_3(X)]^T$$

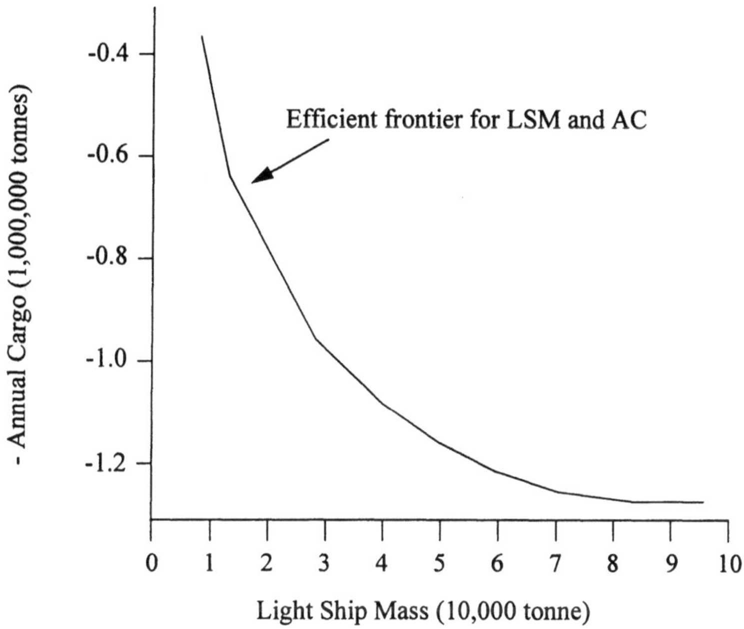

Figure 4.17c *Local Efficient Frontier between LSM* and *AC*

Then, we can calculate the gradient of the three objectives at the point X^0, namely $\nabla f_1(X^0)$, $\nabla f_2(X^0)$ and $\nabla f_3(X^0)$. If the first objective is selected as reference and equal weights are assigned to the three objectives, so that

$$w_{11}^0 = w_{21}^0 = w_{31}^0 = 1$$

we can then formulate problem (4.41). Solving the problem, we get the first efficient ship design X^1 with

$$F(X^1) = [-14.707 \quad -9.3045 \quad 1.2411]^T$$

At X^1, the first interaction can be conducted as follows. First, suppose objective 1 is still selected as the reference objective. At X^1, the following two indifference tradeoffs are provided

$$F(X^1) \sim [-14.707 + 0.1 \quad -9.3045 - 2 \quad 1.2411]^T$$

$$F(X^1) \sim [-14.707 + 0.1 \quad -9.3045 \quad 1.2411 - 0.5]^T$$

Note that $f_3(X)$ is for maximisation. Thus, we have $\Delta_1^1 = 0.1$, $\Delta_2^1 = 2$, $\Delta_3^1 = 0.5$. Hence

$$w_{11}^1 = 1, \ w_{21}^1 = \frac{\Delta_1^1}{\Delta_2^1} = 0.05, \ w_{31}^1 = \frac{\Delta_1^1}{\Delta_3^1} = 0.2$$

Then, the gradients $\nabla f_j(X^1)$ (j=1, 2, 3) can also be calculated and problem (4.41) can be re-constructed at X^1. Solving problem (4.41), we get $\overline{X}^2$ and

$$F(\overline{X}^2) = [-12.81 \quad -0.72 \quad 0.37]^T$$

From the direction $d^1 = (\overline{X}^2 - X^1)$, the utility function will increase. To determine a better solution than X^1 along this direction, Table 4.1 is constructed for X^1 and $\overline{X}^2$, as shown by Table 4.7.

Table 4.7 *Objective Values at Solutions along* $d^1 = (\overline{X}^2 - X^1)$

t^q	0.0	0.2	0.4	0.6	0.8	1.0
$f_1(X^1 + t^q d^1)$	14.7	12.44	10.73	9.72	9.85	12.81
$f_2(X^1 + t^q d^1)$	9.30	6.49	4.29	2.65	1.48	0.72
$f_3(X^1 + t^q d^1)$	1.24	1.17	1.07	0.91	0.67	0.37

Suppose the solution $X^1 + t^q d^1$ at $t^t = 0.4$ has the highest utility among the six solutions as listed in Table 4.7. Then, let $X^2 = X^1 + 0.4d^1$ and we have

$$\Delta^0 = |(W^0)^T[F(X^1) - F(X^0)]| = 2.8735$$

$$\Delta^1 = |(W^1)^T[F(X^2) - F(X^1)]| = 4.1954$$

Thus, $\Delta^1/\Delta^0 = 1.46 > 0.01$. The interaction should continue.

At X^2, objective 1 is still chosen as the reference objective and the following two indifference tradeoffs are provided

$$F(X^2) \sim [-10.7276 + 0.1 \quad -4.2944 - 1.5 \quad 1.0686]^T$$

$$F(X^2) \sim [-10.7276 + 0.1 \quad -4.2944 \quad 1.0686 - 0.1]^T$$

These tradeoffs lead to the assignment of the following weights

$$w_{11}^1 = 1, \quad w_{21}^0 = 0.0667, \quad w_{31}^0 = 1$$

For the second interaction, the gradient $\nabla f_j(X^2)$ (j=1, 2, 3) can be calculated and problem (4.41) can be re-constructed at X^2. Solving the new problem, we get $\overline{X}^3$ and

$$F(\overline{X}^3) = [-12.82 \quad -0.72 \quad 0.37]^T$$

Along the direction $d^2 = \overline{X}^3 - X^2$, the utility function will increase in a neighborhood of X^2. The solutions along d^2 between X^2 and $\overline{X}^3$ are sampled as shown in Table 4.8.

Table 4.8 *Objective Values at Solutions along* $d^2 = (\overline{X}^3 - X^2)$

t^q	0.0	0.2	0.4	0.6	0.8	1.0
$f_1(X^1 + t^q d^1)$	10.73	10.02	9.63	9.69	10.48	12.82
$f_2(X^1 + t^q d^1)$	4.29	3.24	2.38	1.68	1.13	0.72
$f_3(X^1 + t^q d^1)$	1.07	0.98	0.86	0.72	0.55	0.37

Suppose the solution $X^3 = X^2 + 0.2d^2$ has the highest utility among the six solutions listed in Table 4.8, or $F(X^3) = [-10.02, -3.24, 0.98]^T$. We then have

$$\Delta^2 = \left| (W^2)^T [F(X^3) - F(X^2)] \right| = 0.6882$$

We then have $\Delta^2/\Delta^0 = 0.2395 > 0.01$. Thus, the interaction has to be continued further

At X^3, objective 1 is still chosen for reference. The following indifference tradeoffs are provided.

$$F(X^3) \sim [-10.0201 + 0.1 \quad -3.2448 - 1.3 \quad 0.9793]^T$$

$$F(X^3) \sim [-10.0201 + 0.1 \quad -3.2448 \quad 0.9793 - 0.03]^T$$

Then, construct problem (4.41) at X^3. Solving the problem we get $\overline{X}^4$ as shown by the last column of Table 4.9, where $d^3 = \overline{X}^4 - X^3$ and $F(\overline{X}^4) = [-11.05, -0.94, 0.53]^T$.

Table 4.9 *Objective Values at Solutions along* $d^2 = (\overline{X}^3 - X^2)$

t^q	0.0	0.2	0.4	0.6	0.8	1.0
$f_1(X^1 + t^q d^1)$	10.02	9.69	9.53	9.59	10.01	11.05
$f_2(X^1 + t^q d^1)$	3.24	2.63	2.10	1.64	1.25	0.94
$f_3(X^1 + t^q d^1)$	0.98	0.91	0.83	0.74	0.64	0.53

Suppose the solution $X^4 = X^3 + 0.2d^3$ has the highest utility among the six solutions as listed in Table 4.9, or $F(X^4) = [-9.69, -2.63, 0.91]^T$. Then we can have $\Delta^3/\Delta^0 = 0.0158 > 0.01$. Thus, the process needs to proceed.

Again, use objective 1 as the reference objective. Given the following indifference tradeoffs

$$F(X^4) \sim [-9.69 + 0.1 \quad -2.63 - 1.5 \quad 0.91]^T$$

$$F(X^4) \sim [-9.69 + 0.1 \quad -2.63 \quad 0.91 - 0.02]^T$$

we can once again solve problem (4.41) constructed at X^4, resulting in $\overline{X}^5$ with $F(\overline{X}^5) = [-14.11, -8.44, 1.24]^T$, as shown in Table 4.10.

Table 4.10 *Objective Values at Solutions along* $d^2 = (\overline{X}^3 - X^2)$

t^q	0.0	0.2	0.4	0.6	0.8	1.0
$f_1(X^1 + t^q d^1)$	9.69	10.13	10.84	11.76	12.85	14.11
$f_2(X^1 + t^q d^1)$	2.63	3.47	4.46	5.61	6.93	8.44
$f_3(X^1 + t^q d^1)$	0.91	1.02	1.10	1.16	1.20	1.24

Suppose $X^5 = X^4 + 0.2d^4$ has the highest utility among the six solutions as listed in Table 4.10. We can have $\Delta^4/\Delta^0 = 0.0016 < 0.01$. X^5 is thus determined as the best compromise solution. The values of the design variables and objectives at X^5 are given by

$$X^5 = [340.54 \quad 18.68 \quad 25.68 \quad 0.63 \quad 49.86 \quad 14.00]^T$$

$$f_1(X^5) = 10.1321, \quad f_2(X^5) = 3.4675, \quad f_3(X^5) = 1.0154$$

However, X^5 is not an efficient solution as there exist other solutions which dominate X^5. For instance, we can generate the following solution X^6 using the ISTM method

$$X^6 = [342.34 \quad 19.38 \quad 26.69 \quad 0.6362 \quad 54.05 \quad 14.00]^T$$

$$f_1(X^6) = 10.1318, \quad f_2(X^6) = 3.4671, \quad f_3(X^6) = 1.0412$$

Obviously $f_1(X^6) = f_1(X^5)$ and $f_2(X^6) = f_2(X^5)$ but $f_3(X^6) > f_3(X^5)$. In other words, designs X^6 and X^5 have the same achievement levels on the *transportation cost* (f_1) and *light ship mass* (f_2), but the former has 25800 tonnes of extra *annual cargo carrying capacity*. However, X^5 may still be quite a good design as the relative difference between $f_3(X^6)$ and $f_3(X^5)$ is only about 2.54%.

By using the ISTM method, we can also generate ship designs in an interactive manner. The solutions generated using the ISTM method and their achievement levels on the objectives are shown in Table 4.11. Note that the three extreme efficient solutions listed in Table 4.6 are among the ten solutions, that is, $\tilde{X}^1=\overline{X}^1$, $\tilde{X}^4=\overline{X}^2$ and $\tilde{X}^{10}=\overline{X}^3$. The preference relations between the generated solutions are provided as follows on the basis of the achievement levels of each design with respect to not only the three objectives but also the value of other intermediate iterms such as *deadweight* (DW) and the design variables.

$$\Omega_p = \left\{ (\tilde{X}^l \quad \tilde{X}^h) \left| \begin{array}{l} \tilde{X}^1\mathbf{P}\tilde{X}^2,\ \tilde{X}^3\mathbf{P}\tilde{X}^2,\ \tilde{X}^1\mathbf{P}\tilde{X}^3,\ \tilde{X}^3\mathbf{P}\tilde{X}^4,\ \tilde{X}^3\mathbf{P}\tilde{X}^5,\ \tilde{X}^5\mathbf{P}\tilde{X}^4 \\ \tilde{X}^1\mathbf{P}\tilde{X}^6,\ \tilde{X}^5\mathbf{P}\tilde{X}^6,\ \tilde{X}^6\mathbf{P}\tilde{X}^7,\ \tilde{X}^7\mathbf{P}\tilde{X}^8,\ \tilde{X}^8\mathbf{P}\tilde{X}^9,\ \tilde{X}^9\mathbf{P}\tilde{X}^{10} \end{array} \right. \right\} \tag{4.89}$$

Table 4.11 *Achievement Levels of Generated Solutions*

efficient designs	*TC* (*pounds/tonne*)	*LSM* (/10,000 *tonnes*)	*AC* (/1,000,000 *tonnes*)
$\tilde{X}^1$	9.458679	2.235511	0.888580
$\tilde{X}^2$	10.979374	1.616371	0.769986
$\tilde{X}^3$	11.752936	2.514088	0.960127
$\tilde{X}^4$	12.814046	0.716313	0.371948
$\tilde{X}^5$	13.735794	5.012039	1.173279
$\tilde{X}^6$	14.536130	5.912211	1.209051
$\tilde{X}^7$	15.336290	6.812314	1.235583
$\tilde{X}^8$	16.136290	7.712314	1.255784
$\tilde{X}^9$	16.946248	8.612650	1.268370
$\tilde{X}^{10}$	17.341306	9.614508	1.270249

In addition to the 12 pairwise comparisons above of the generated efficient solutions given by equation (4.89), three different sets of relative weights are also provided. Let w_1, w_2 and w_3 denote the weights of $f_1(X)$, $f_2(X)$ and $f_3(X)$, respectively. The first set of weights is given by

$$\frac{w_1}{w_2} = 3, \qquad \frac{w_1}{w_3} = 1.5 \tag{4.90}$$

To examine the impact of the variations of the weights on the final solutions, the following two sets of weights are also examined

$$\frac{w_1}{w_2} = 2, \qquad \frac{w_1}{w_3} = 1 \tag{4.91}$$

$$\frac{w_1}{w_2} = 2, \qquad \frac{w_1}{w_3} = 1.5 \tag{4.92}$$

where *transportation cost* is always regarded as the most important objective and *light ship mass* as the least important with different degrees of intensity. Note that $\Omega_f = \{(1, 2); (1, 3)\}$ for all the three sets of weights. Given the unlimited cargo availability, economy of scale would therefore tend to favour relatively large vessels subject to any draft restrictions that may exist.

4.4.4 Design by Setting Target Values

Given the relative weights of the three objectives in equations (4.90) to (4.92), the best compromise ship design can be sought by setting target values for the three objectives on the basis of the pay-off table of the ship design problem as shown by Table 4.6.

If the target values are taken from the pay-off table to define the following ideal design F^-

$$F^- = [f_1(\overline{X}^1) \quad f_2(\overline{X}^2) \quad f_3(\overline{X}^3)]^T$$

we can use the minimax method to obtain compromise designs which are nearest to the defined ideal design in the sense of the weighted ∞-norm. With the three sets of weights for the three objectives, as given by equations (4.90), (4.91) and (4.92), following the steps of the minimax method as described in section 4.3.3 we can generate three compromise designs, denoted by $\hat{X}^1$, $\hat{X}^2$ and $\hat{X}^3$, respectively. The values of the design variables and objectives of these designs are as shown by Table 4.12.

From Table 4.12, one can find that design $\hat{X}^1$ has achieved lowest *transportation cost* among the three designs. This is because the set of weights provided by equation (4.90) gives the highest weight to $f_1(X)$. Similarly, $\hat{X}^3$ has achieved the

Table 4.12 *Compromise Ship Designs*

extreme designs	variables						objectives		
	x_1	x_2	x_3	x_4	x_5	x_6	$f_1(X)$	$f_2(X)$	$f_3(X)$
$\hat{X}^1$	345.6	20.5	28.2	0.6	57.6	14.0	10.54	4.07	1.09
$\hat{X}^2$	335.4	20.9	28.9	0.7	55.9	14.0	10.94	4.05	1.10
$\hat{X}^3$	314.7	19.9	27.4	0.7	52.5	14.0	10.68	3.48	1.06

lowest *light ship mass* and $\hat{X}^2$ the highest *annual cargo carrying ability*. These achievements are also in harmony with the assignment of weights given by equations (4.90) to (4.92). The choice of the best compromise design depends upon which set of weights can best represent the designer's priorities.

Instead of defining an ideal design from the pay-off table, the designer is free to assign a set of target values for the objectives. For instance, the designer may assign the following goal values for the three objectives

$$\hat{F} = [\hat{f}_1 \quad \hat{f}_2 \quad \hat{f}_3]^T = [9.5 \quad 2.0 \quad 1.0]^T$$

Suppose a design is expected to achieve the objective values as close as to the assigned target values as possible. Suppose the relative weights of the objectives are given by equations (4.90) to (4.92), respectively, with no objectives being absolutely more important than others. Using goal programming as discussed in section 4.3.1, we can obtain three compromise designs, denoted by $\hat{X}^1$, $\hat{X}^2$ and $\hat{X}^3$ corresponding to the assigned three sets of weights. The results are as shown by Table 4.13.

Table 4.13 *Compromise Ship Designs Obtained Using Goal Programming*

extreme designs	variables						objectives		
	x_1	x_2	x_3	x_4	x_5	x_6	$f_1(X)$	$f_2(X)$	$f_3(X)$
$\hat{X}^1$	289.65	17.37	23.82	0.63	48.28	14.00	9.569	2.558	0.938
$\hat{X}^2$	300.18	17.96	24.65	0.63	50.03	14.00	9.690	2.809	0.971
$\hat{X}^3$	284.94	17.11	23.45	0.63	47.49	14.00	9.525	2.451	0.923

It can be seen from Table 4.13 that $\hat{X}^1$, $\hat{X}^2$ and $\hat{X}^3$ are not significantly different from one another although the target value of 9.5 for the first objective $f_1(X)$ is always better attained than the other objectives. This is because $f_1(X)$ is always regarded as the most important objective.

4.4.5 Adaptive and Compromise Design

It may be noted that the best solution among the 10 solutions generated using the ISTM method, as shown in Table 4.11, is $\tilde{X}^1$ which is also the extreme solution minimising $f_1(X)$. However, the 10 solutions are not entirely ranked by $f_1(X)$ as $\tilde{X}^3 \mathbf{P} \tilde{X}^2$ and $\tilde{X}^5 \mathbf{P} \tilde{X}^4$. Piecewise linear utility functions may then be estimated using the local utility function method to search for the best compromise solution.

From Table 4.6, the acceptable intervals of the three objectives are given by

$$[f_1^0 \quad f_1^*] = [17.3413 \quad 9.4587]$$

$$[f_2^0 \quad f_2^*] = [9.6145 \quad 0.7163]$$

$$[f_3^0 \quad f_3^*] = [0.8886 \quad 1.2702]$$

Let $f_i^- = f_i^0$ (i=1, 2, 3). Suppose each of the three intervals is cut into five equal sub-intervals or N_i=5 (i=1, 2, 3). Then the values of the end points of the equal intervals are given by Table 4.14 (see equation (4.61)).

Based on the information given in the last subsection and the weight assignment given by equations (4.90), a linear goal programming estimation model as defined by problem (4.64) can be constructed, which is composed of 54 variables, including 18 marginal utilities $u_i(f_i^j)$ (j=0, 1, $\cdots$, 5; i=1, 2, 3), 20 approximation error variables σ_l^+ and σ_l^- for l=1, $\cdots$, 10 and 16 inconsistency error variables σ_{lh}^+ for $(\tilde{X}^l, \tilde{X}^h) \in \Omega_p$ and s_{ij} for $(i, j) \in \Omega_f$, and 45 linear/nonlinear constraints, including 27 linear inequality constraints, 6 linear equality constraints and 12 nonlinear equality constraints.

This linear goal programming problem has infinite number of optimal solutions with all error variables being zero. Problems (4.66) and (4.67) are then constructed and solved. As a result, 36 optimal estimates for each marginal utility, say u_i^j, are sampled. The mean of these 36 estimates is denoted by $\hat{u}_i^j$ and given in Table 4.14. The piecewise linear marginal utility functions are defined by equation (4.65) and illustrated by Figure 4.18. The utility of a solution may then be calculated using Table 4.14 and equation (4.60).

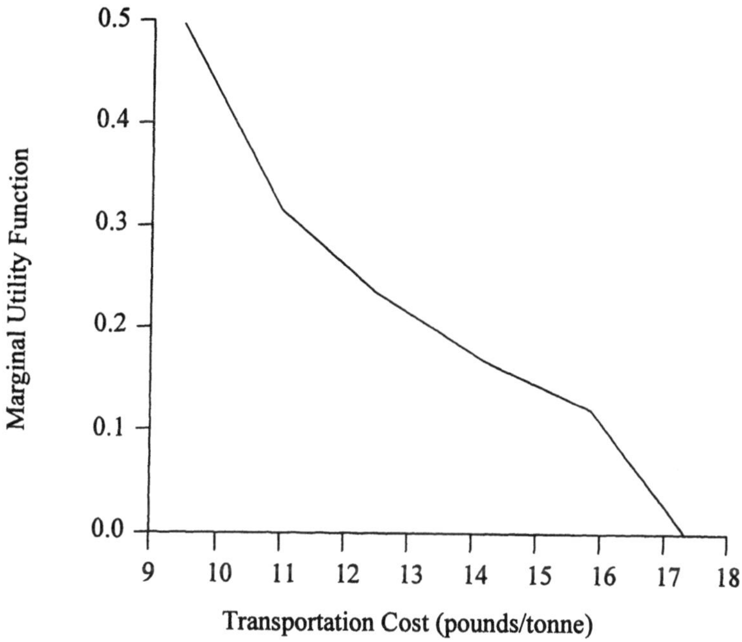

Figure 4.18a *Marginal Utility Function of TC*

The best compromise solution may be generated by solving problem (4.68). The coefficients α_{ij}, β_i and γ_i as defined by equations (4.69) and (4.70) are obtained as follows

$$[\alpha_{11} \;\; \alpha_{12} \;\; \alpha_{13} \;\; \alpha_{14}] = [-0.021916 \;\; 0.005454 \;\; 0.005044 \;\; 0.027751]$$

$$[\alpha_{21} \;\; \alpha_{22} \;\; \alpha_{23} \;\; \alpha_{24}] = [-0.012363 \;\; -0.011743 \;\; -0.001601 \;\; 0.005759]$$

$$[\alpha_{31} \;\; \alpha_{32} \;\; \alpha_{33} \;\; \alpha_{34}] = [-0.112688 \;\; 0.007059 \;\; -0.070777 \;\; 0.043494]$$

$$[\beta_1 \;\; \beta_2 \;\; \beta_3] = [0.093149 \;\; 0.031636 \;\; 0.449170]$$

$$[\gamma_1 \;\; \gamma_2 \;\; \gamma_3] = [1.433826 \;\; 0.335516 \;\; -0.142444]$$

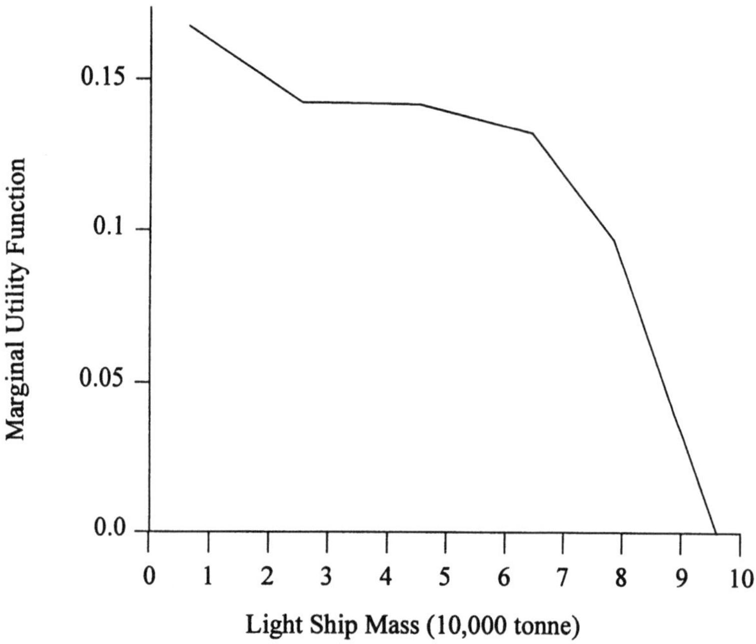

Figure 4.18b *Marginal Utility Function of LSM*

Given the above coefficients and with Ω being defined by equation (4.88) and $f_1(X)$, $f_2(X)$ and $f_3(X)$ by equations (4.75) to (4.77), problem (4.68) can then be solved, resulting in the best compromise design, defined by $\tilde{X}^{11}$ and shown by Table 4.15.

The utilities of the 11 generated efficient solutions are thus obtained as follows

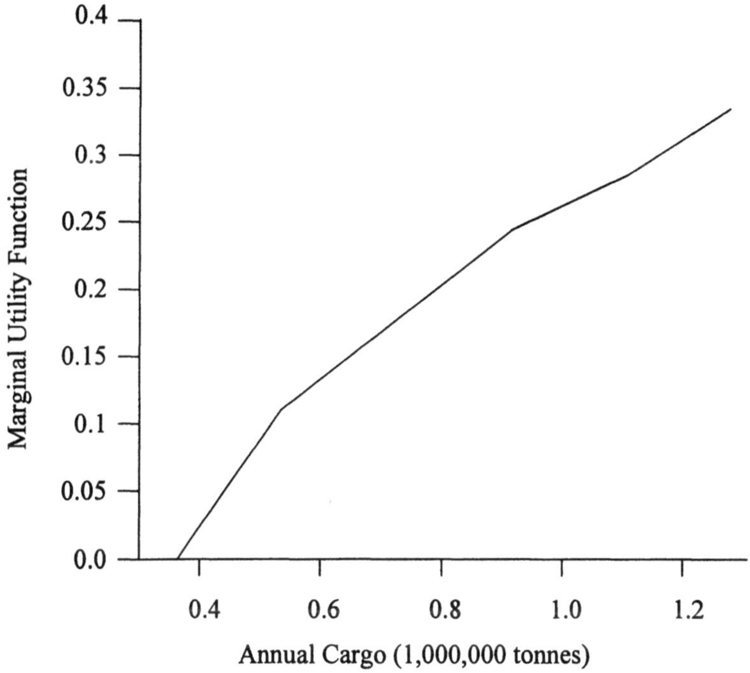

Figure 4.18c *Marginal Utility Function of AC*

$$[u(\tilde{X}^{11}) \quad u(\tilde{X}^{1}) \quad u(\tilde{X}^{2}) \quad u(\tilde{X}^{3}) \quad u(\tilde{X}^{4}) \quad [u(\tilde{X}^{5})$$
$$u(\tilde{X}^{6}) \quad u(\tilde{X}^{7}) \quad u(\tilde{X}^{8}) \quad u(\tilde{X}^{9}) \quad u(\tilde{X}^{10})]$$
$$= [0.8784 \quad 0.8760 \quad 0.6727 \quad 0.6811 \quad 0.4001 \quad 0.6387$$
$$0.6157 \quad 0.5769 \quad 0.5164 \quad 0.4147 \quad 0.3333]$$

The utility of $\tilde{X}^{11}$ is slightly better than $\tilde{X}^{1}$. $\tilde{X}^{11}$ is more favourable than $\tilde{X}^{1}$ in that the former has almost the same *transportation cost* as the latter but slightly higher *annual cargo* which is important.

Table 4.14 *Estimated Mean Marginal Utilities*

end points	*TC*		*LSM/10,000*		*AC/1,000,000*	
j	f_1^j	$\hat{u}_1^j$	f_2^j	$\hat{u}_2^j$	f_3^j	$\hat{u}_3^j$
0	17.341	0.0000	9.615	0.0000	0.372	0.0000
1	15.765	0.1211	7.835	0.0918	0.552	0.1046
2	14.188	0.1731	6.055	0.1396	0.731	0.1687
3	12.612	0.2423	4.276	0.1456	0.911	0.2353
4	11.035	0.3274	2.496	0.1459	1.091	0.2765
5	9.459	0.5000	0.716	0.1667	1.270	0.3333

Table 4.15 *Best Compromise Designs*

best	variables						objectives		
designs	x_1	x_2	x_3	x_4	x_5	x_6	$f_1(X)$	$f_2(X)$	$f_3(X)$
$\tilde{X}^{11}$	281.42	16.91	23.16	0.63	46.90	14.00	9.498	2.373	0.911
$\tilde{X}^{12}$	301.79	18.04	24.78	0.63	50.30	14.00	9.711	2.848	0.976
$\tilde{X}^{13}$	283.28	17.02	23.31	0.63	47.21	14.00	9.512	2.414	0.917

To test the impact of the changes of the weights on the best solutions, two different sets of weights as given by equations (4.91) and (4.92) are used to estimate utility functions. Figure 4.19 shows the estimated marginal utility functions where u_{i1}, u_{i2} and u_{i3} denote the marginal utility functions of $f_i(X)$ estimated using the different sets of weights given by (4.90), (4.91) and (4.92), respectively. Optimising the two new utility functions within Ω results in two new compromise solutions denoted by $\tilde{X}^{12}$ and $\tilde{X}^{13}$. The objective achievement levels and the values of the design variables at $\tilde{X}^{12}$ and $\tilde{X}^{13}$ are as shown in Table 4.15.

From Table 4.15, it can be seen that $\tilde{X}^{11}$ and $\tilde{X}^{13}$ are not significantly different from each other. This is because the weights as given by equations (4.90) and (4.92) are only slightly different and the utility functions u_{i1} and u_{i3} (i=1, 2, 3) are similar as shown by the solid and broken lines in Figure 4.19. $\tilde{X}^{12}$ is slightly different from $\tilde{X}^{11}$ in that $\tilde{X}^{12}$ has higher *annual cargo* and also a somewhat higher *transportation cost* and *light ship mass* while the marginal utility function u_{32} increases faster than the decrease of u_{12} and u_{22} in a neighbourhood of $\tilde{X}^{12}$ as shown by the doted lines in Figure 4.19. On the whole, it may be concluded that the three compromise solutions are not significantly different from

one another and that the best compromise solution may be either $\tilde{X}^{11}$ or $\tilde{X}^{12}$ or $\tilde{X}^{13}$, depending upon which set of the weights may be ultimately employed by the DM.

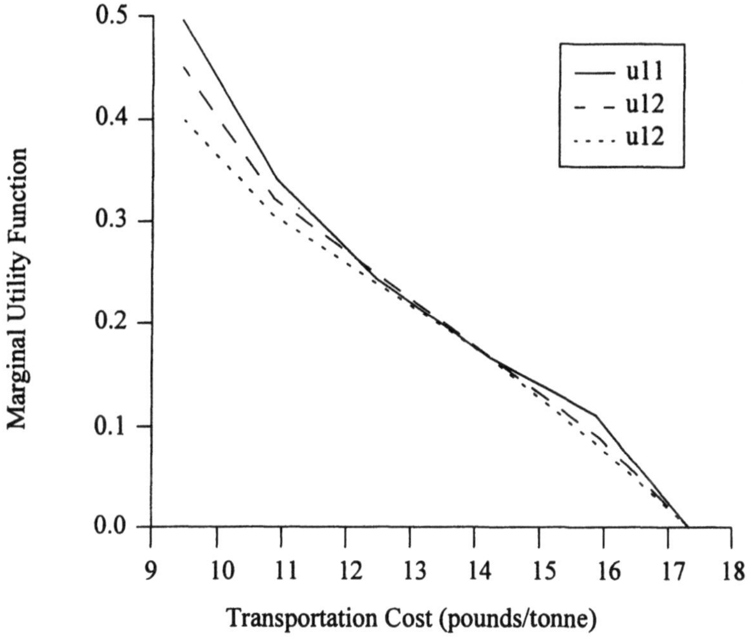

Figure 4.19a *Three Marginal Utility Function of TC*

4.5 Concluding Comments

The aim of MODM is to assist in the synthesis of technical alternatives. Such synthesis inevitably involves conflict and hence trade-off, and it is often

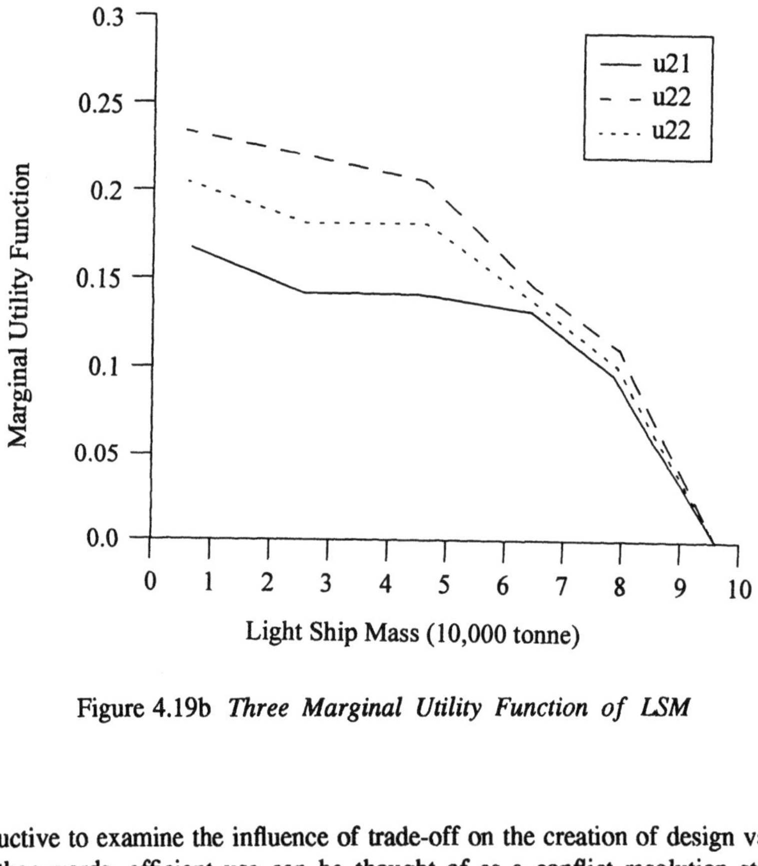

Figure 4.19b *Three Marginal Utility Function of LSM*

instructive to examine the influence of trade-off on the creation of design variety. In other words, efficient use can be thought of as a conflict resolution strategy that creates the variety that is one of the essential features of design. In the process of pursuing the above reconciliation the DM or designer gets to know more about his own sense of priorities and this too has a beneficial effect on the resulting decision making.

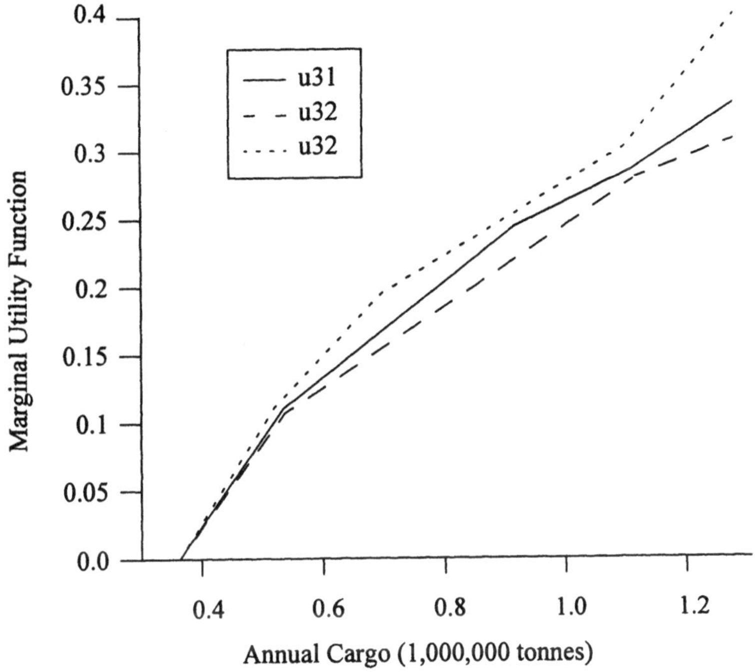

Figure 4.19c *Three Marginal Utility Function of AC*

However, not every design or decision making problem comes neatly packaged in terms of the analytical demands it makes. What is often required is a computational tool that can negotiate complex, often multi-modal functions. This is where evolutionary computational tools prove their worth. Chapter 5 examines one such approach in greater detail in the context of MCDM

5
Multiple Criteria Decision Making and Genetic Algorithms

5.1 Introduction

In all of the decision-making methods that have been considered so far, the implicit assumption is that the computational aspects of the problem can be handled satisfactorily. This is often a reasonable assumption, but not necessarily so. This is because the search for the "best solution" in real life in the presence of multiple criteria or measures of performance is often significantly and indeed critically influenced by the ability to assess the effect of design changes on system performance. Difficulties may arise because the mathematical models used for such purposes are possibly discontinuous. Moreover, the performance landscape may not necessarily be single-peaked (or unimodal). This essentially means that one needs a robust optimisation method that can cope with noisy, multi-peaked performance relations with discontinuities in them, in recognition of the fact that real life problems often present themselves in these inconvenient forms.

In situations like these, many of the adaptive computational tools become attractive. Adaptive algorithms, as their name implies, mimic some aspects of the behaviour of natural biological systems. In the main this consists of random experimentation combined with search for efficiency or fitness. With the increased computing power at our disposal nowadays it is often economic to take advantage of wide but controlled numerical experimentation even for very complex applications in an attempt to ensure that the entire solution space is thoroughly investigated, such that all pockets of efficiency are identified, if possible.

One particular algorithm which demonstrates this adaptive or evolutionary nature of exploration of the solution space is the Genetic Algorithm (GA). A genetic algorithm is a search technique based on the Darwinian principles of natural selection. A basic feature of this approach is to represent the set of variables defining a problem as a numerical string, either binary or otherwise. Starting with a pool of solutions (or strings) the fittest members are chosen for the next generation and off-springs are created by crossover and mutation operators acting on the strings. Crossover results in the exchange of genetic material or features between relatively fit members of the population leading potentially to a better pool of solutions. Mutation introduces random changes to the gene pool in an attempt to ensure that the exploration aspect of the solution strategy is not prematurely concluded in the drive for convergence. The presence of multiple, potentially conflicting criteria makes the solution process somewhat more elaborate but the guiding principles remain unchanged.

In order to fully examine the multiple criteria genetic algorithm it is helpful to examine the single criterion version first, and this is what is examined next.

5.2 The Mechanics of the Simple Genetic Algorithm

A small problem, one of maximising the value of an 8-bit binary string is used to demonstrate the operation of a GA.

Two characteristic features of GA are that

(i) GAs search on the basis of a population of solutions.
(ii) GAs use a range of probabilistic rules, as opposed to deterministic transition rules.

5.2.1 Selection, Crossover and Mutation

The GA search consists of the following steps:

<u>Step 1</u>: Initial Population

This is the starting event in all GA searches. As GAs work with populations of solutions rather than individual ones, it is necessary to define the initial population. Fortunately, this initial population can consist of any collection of solutions that is reasonably representative of the whole solution space. This latter condition is important as an initial population that is only representative of a small part of the solution space can lead to premature convergence to a local minimum. The initial population can be conveniently generated by a random sampling over the whole solution space. The population size, however, is very much dependent on the problem being tackled and the type of GA being used. The principal requirements are those of exploration and convergence. In other words the population should be large enough to allow a reasonably representative sample of candidate solutions to be present while not being so large that it hampers the convergence of the solution.

In this simple example a population size of 6 is chosen, as follows, using a binary representation, for illustrative purposes only.

Initial Population

0	1	0	0	1	0	0	0	72	String 1
1	0	1	1	1	0	1	0	186	String 2
0	0	0	0	0	1	0	1	5	String 3
0	1	0	0	0	0	0	1	65	String 4
1	1	0	1	0	1	1	0	214	String 5
0	0	0	1	1	0	0	0	24	String 6

Average Fitness = 94.3
Best Design Fitness = 214

Step 2: Selection

There are several methods of selecting candidates that survive from one generation to the next. In this example a Roulette Wheel selection is used [Figure 5.1]. This is a fairly common selection procedure where each String is allocated a segment on a notional roulette wheel. The size of the segment is proportional to the "fitness" of the corresponding string, represented here by the magnitude of the number represented by the string.

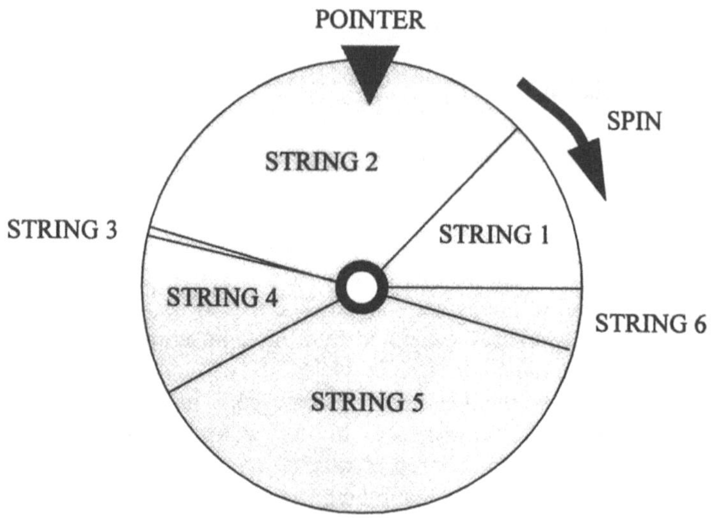

Figure 5.1 *Roulette Wheel Selection*

The wheel is then "spun" numerically and the string corresponding to the segment in which the pointer lands is chosen to join the selected population. As String 5 has the highest fitness, corresponding to nearly 38% of the disc it maybe expected that more than one copy of String 5 will be sampled for a six member population.

As the selected population below shows, three copies of String 5 are chosen for the selected population in this experiment. Those strings that are not chosen for this population die out. (Strings 1 and 3). They may appear again in later generations because of the action of genetic operators on other strings (see Step 3 below).

The selected population obtained is as follows:

1 1 0 1 0 1 1 0	214	String 1 – was String 5
1 0 1 1 1 0 1 0	186	String 2 – was String 2
0 0 0 1 1 0 0 0	24	String 3 – was String 6
1 1 0 1 0 1 1 0	214	String 4 – was String 5
0 1 0 0 0 0 0 1	65	String 5 – was String 4
1 1 0 1 0 1 1 0	214	String 6 – was String 5

Step 3: Crossover and Mutation

The selected population at the end of Step 2 above forms the mating pool. They are acted upon by the crossover and mutation operators.

Crossover:

There are many different kinds of crossover operator. The simplest of these is the single point crossover. In this approach pairs of strings are randomly chosen. Crossover is then performed on these strings with a stipulated likelihood of Pc (usually 0.6 – 0.9). Crossover consists of selection of a random crossing point for these pairs of strings and swapping over their tails to generate two new offspring strings, as shown below. Each string in the mating process is only paired up once.

Cross Strings 1 and 5 at Position 4

1 1 0 1 | 0 1 1 0	→	1 1 0 1 0 0 0 1	String 1
0 1 0 0 | 0 0 0 1		0 1 0 0 0 1 1 0	String 5

Cross Strings 3 and 4 at Position 6

0 0 0 1 1 0 | 0 0	→	0 0 0 1 1 0 1 0	String 3
1 1 0 1 0 1 | 1 0		1 1 0 1 0 1 0 0	String 4

Strings 2 and 6 are left untouched.

Mutation:

Mutation is used to introduce small random changes in the population with a small probability P_m (usually 0.001-0.01). In this process a bit in a strong is chosen at random and its current state is switched, as shown below. The higher the probability of mutation the more the search process functions like a pure random search.

Mutate String 2 Position 2

1 0 1 1 1 0 1 0 → 1 1 1 1 1 0 1 0 String 2

Mutate String 5 Position 7

0 1 0 0 0 1 1 0 → 0 1 0 0 0 1 0 0 String 5

Strings 1,3,4 and 6 Left Untouched

The resultant new population obtained is as shown below. It is clear that both the average fitness and the fitness of the fittest member has increased, the former going up from 94.3 to 159.3 and the latter from 214 to 250. The GA has thus made significant improvements in fitness in one generation. As may be expected this improvement in fitness tapers off in moving from generation to generation, as most of the realisable gains are obtained early on.

New Population

1	1	0	1	0	0	0	1	186	String 1
1	1	1	1	1	0	1	0	250	String 2
0	0	0	1	1	0	1	0	26	String 3
1	1	0	1	0	1	0	0	212	String 4
0	1	0	0	0	1	0	0	68	String 5
1	1	0	1	0	1	1	0	214	String 6

Average Fitness = 159.3
Best Design Fitness = 250

Step 4: Return to Step 2 or STOP

As the improvement in fitness falls with each passing generation it is obvious that the process converges asymptotically. It is thus necessary to stop the search on the basis of some convergence criteria. These are usually based on the improvement in average fitness or the fitness of the fittest member and when improvements in one or both fall below certain threshold values the search is terminated. Otherwise members of a new population are chosen according to the established procedure starting from Step 2.

Figure 5.2 shows the convergence behaviour of the example problem. It is clear that the fitness approaches the number 255 as this is the largest number capable of being represented by a 8-bit binary string. The genetic algorithm finds the solution of maximum fitness rather easily for this simple problem.

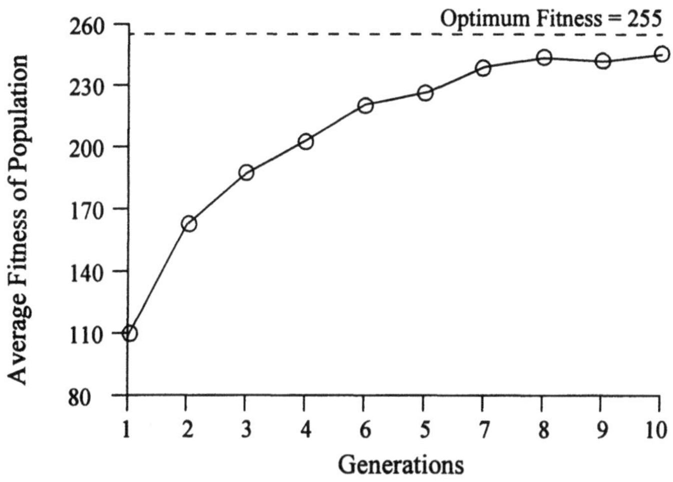

Figure 5.2 *Convergence of GA search example*

5.2.2 A Bi-Modal Optimisation Problem.

As indicated earlier, GAs are designed to cope with noisy functions. Such functions have multiple fitness peaks and the aim of the GA is to find the highest point of the highest peak. An example function in a single variable x is the bi-modal function

$$f(x) = 4 - \left[\left(\frac{x}{1.6} - 0.5\right)\left(\frac{x}{1.6} - 3\right)\left(\frac{x}{1.6} - 6.2\right)\left(\frac{x}{1.6} - 2\right) / 6.0\right]$$

The function is plotted in Figure 5.3 for $0 \le x \le 10$.

It is obvious that a hill-climbing algorithm would identify one of the two hill tops depending on the starting point of the search. A starting value of $x > 4$ would find the true optimum. Hill-climbing algorithms usually address multi-modality by taking a range of start points for the search. The best of the local optima is then taken as the final solution but there is, of course, no guarantee that the best solution would always be obtained. As GAs work naturally with populations of solutions there is the possibility of devising a search strategy that finds the true optimum. This is because some of the members of the initial population are likely to be near the taller of the two peaks and thus will converge to the top of that peak. As always, there is some experimentation involved in finding the best population size and the number of generations as the former determines variety or spread of population characteristics and the latter governs convergence.

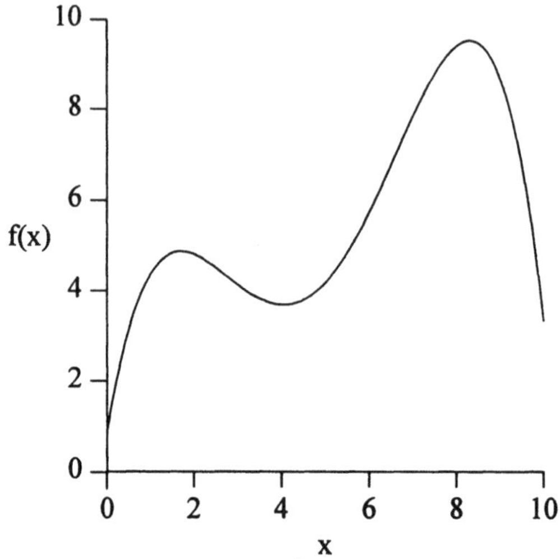

Figure 5.3 *Bi-Modal Test Function*

For this simple bi-modal example a population size of 20 shows a rapid convergence to the solution of interest by generation 20. Using a resolution of 4096 points representing x between 0 and 10 it is possible to find the optimal solution in 20 x 20 = 400 evaluations – or 10% of the possible values. This economy of effort is another reason why GAs are not considered to be as computationally wasteful as may appear at first sight.

Figure 5.4 shows the migration of the GA population with the number of generations. As may be expected most of the movement towards the optimum occurs in the first few generations.

Given the natural analogy between aspects of genetics and GAs, it is helpful to define some of the terminology in genetic terms as shown in Table 5.1 [South 1993].

Table 5.1 – *GA Terminology*

Natural Genetics	Genetic Algorithms
Chromosome	String
Gene	Feature or bit
Allele	Feature or bit value
Genotype	Structure of an individual in the GA population
Phenotype	Decoded individual giving a possible solution to the problem

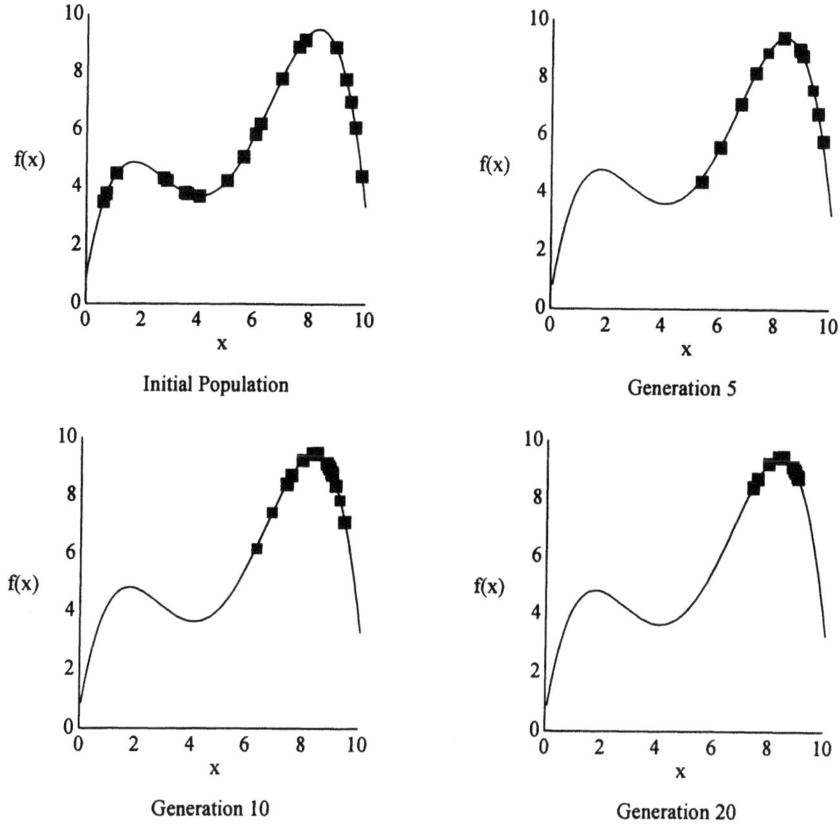

Figure 5.4 *Migration of the GA population for the bi-modal function*

5.2.3 The Need for a Multiple Criteria Approach.

From the discussions above it is clear that the GA approach can be simply represented as an ordered procedure consisting of selection, crossover and mutation operations, applied to each generation of solutions. This can be visualised as shown in Figure 5.5.

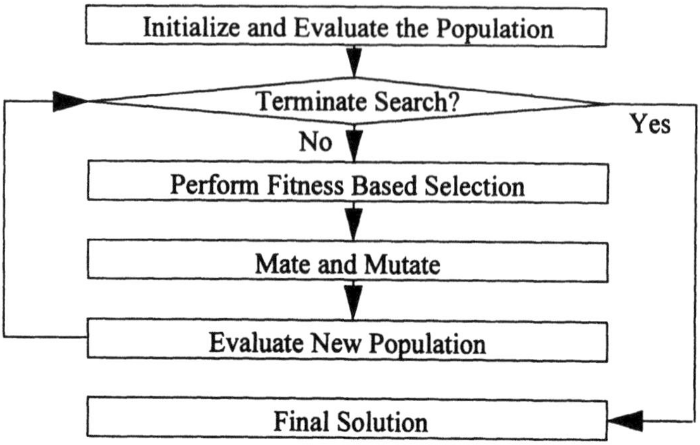

Figure 5.5 *The Basic Genetic Algorithm*

The examples considered thus far have been unconstrained mathematical functions. Most real-life engineering design problems, however, would be constrained by physical laws and functional requirements. Such requirements might also be in mutual conflict so that satisfying all of the requirements simultaneously may not be possible. This is because it is generally not possible to have the best of all possible performance criteria embodied in a single solution. This is clearly the domain of multiple criteria decision making examined upto now. What is required, therefore, is a combination of multiple criteria decision making and genetic algorithms to address those design problems with noisy, multi-modal, possibly discontinuous and potentially conflicting performance requirements.

5.3 Multiple Criteria Genetic Algorithms

The previous sections in this book have argued that many engineering design problems deal, in general, with decisions involving multiple criteria. The concept of applying evolutionary computation tools like GAs to multiple criteria problems is a natural extension. Although this is a relatively new area of investigation several approaches already exist in the literature.

Applying evolutionary computation methods or genetic algorithms to solve multiple criteria decision making problems in engineering design have to deal with the twin issues of searching large and complex solution spaces and dealing with multiple, potentially conflicting objectives. A separate treatment for each of these two objectives has led to two distinct groups of strategies as Figure 5.6 shows.

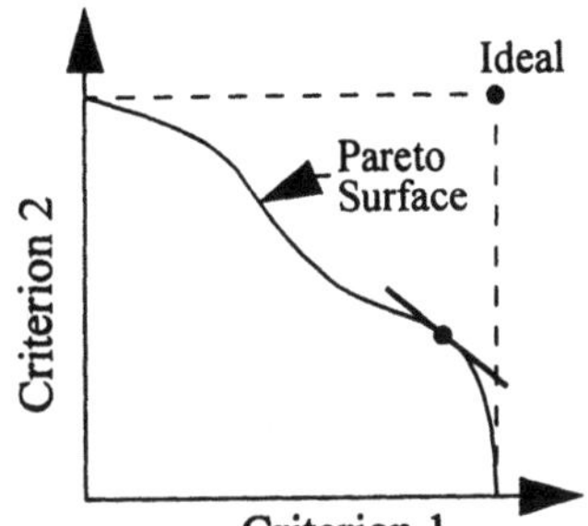

Figure 5.6(a) *Search on the basis of linearly combined criteria*

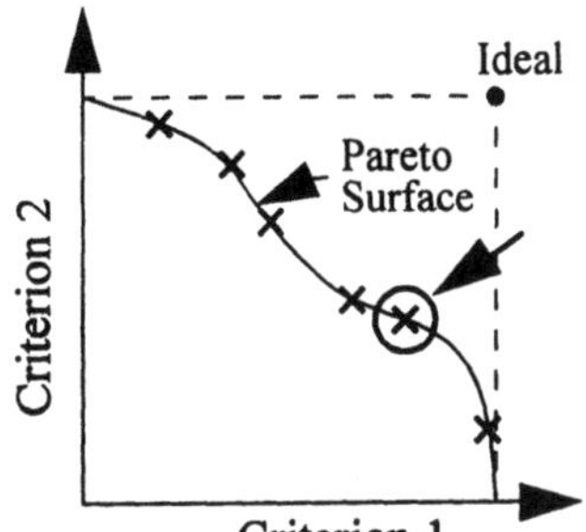

Figure 5.6(b) *Selection from a list of efficient solutions*

These strategies may be described as:

(i) make the multiple criteria decisions first to arrive at a composite measure of fitness by combining the different criteria, and then use the composite measure to search for the best solution. Figure 5.6(a) shows this for a linear combination of criteria. Each linear combination of the two criteria define a line and the point of tangency of this line with the Pareto surface is the "best" solution for this specific combination of criteria.

(ii) conduct the search to assemble a range of possible solutions and then select one or more of these on the basis of multiple criteria decision making. Figure 5.6(b) shows how the best solution from a Pareto Set (marked by crosses) can be selected using some measure of distance from the ideal solution.

5.3.1 Some Comparative Multiple Criteria GA Approaches

The two strategies mentioned above have often been treated independently. In order to simplify the problem further search techniques, including GAs, have often assumed single objectives whereas many multiple objective decision making problems have dealt with a small number of alternatives. This has obviously simplified the task for both approaches but at the expense of necessary numerical adjustments. More recently, however, Pareto-based genetic algorithms have been used to combine search and decision making in a particularly helpful way. Such approaches have used GAs to sample the trade-off surface and applied multiple criteria decision making to narrow the search into areas of interest. The difference between the various approaches is best examined by looking at a few representative approaches.

(a) Multiple Criteria Decisions followed by Search

The commonest approach in this area is to combine the various criteria into a form of additive linear utility function. This utility function is then treated as a fitness function in the GA. The linear approach can be replaced by non-linear terms, for example, by multiplying instead of adding the contributions from attributes but the processing is difficult to control.

The Vector Evaluated Genetic Algorithm (VEGA) [Schaffer 1985] was the first attempt made at extending the GA into the multiple objective domain. Schaffer used a special selection mechanism which chose k equally sized subgroups of individuals from the population based on their performance in each of k criteria. These subgroups were then shuffled together and genetic operators applied. It was recognised that this would favour solutions with extreme performance in at least one objective. To combat this Schaffer suggested applying fitness penalties to locally dominated points and redistributing the deducted fitnesses to non-dominated ones. He found that this caused premature convergence because in populations with few non-dominated points these points were given large fitness values. It was also suggested that individuals performing well in one criterion should be mated with individuals performing well in others. Unfortunately this was found to have a detrimental effect. For this reason random mating was used throughout Schaffer's experiments.

Later analysis of VEGAs performance showed that fitness was effectively a linear combination of criteria where the weights in this linear combination were defined by the distribution of the population [Richardson 1989]. Due to the nature of selection, which was biased towards strings which were strong in at least one criteria, the population contained many extreme individuals and few compromise ones.

Other fitness functions that may be used include the TOPSIS algorithm and the minimax criteria referred to earlier in Sections 3.3.3 and 3.2.4.2.

(b) Search Followed by Multiple Criteria Decision Making.

The aggregation of individual criteria into a single measure can be criticised for being unduly simplistic, as multiple criteria decisions precede a search in this approach. As the information from this search may be vital for the decision making, it makes sense to discover what is attainable in the form of solutions before making any decisions.

The set of obtainable "best" solutions obviously constitutes the Pareto frontier or where solutions on the surface dominate all solutions internal to the surface. The task, therefore, becomes one of finding the nature of the Pareto surface before making decisions.

An approach of this type is the independent sampling method. Using this method a front is built up using a series of independent runs with the weightings between criteria being varied. Fourman [Fourman 1985] used several composite formulae to sample the Pareto surface. Several others have employed similar methods using linear combinations of criteria. However such combinations tend to favour convex parts of the Pareto trade-off surface, because of the linear aggregative nature of the methods.

Rather than use multiple independent composite objective searches as above, some studies have employed parallel searches for members of a single large population. Schaffer was one of the earliest to do so. In his Vector Evaluated Genetic Algorithm (VEGA) 1/k of each new generation is obtained using one of the k criteria. Others like Hajela [Hajela 1992] and Kursawe [Kursawe 1991] use related approaches. All such approaches have been criticised for their bias against individuals not excelling in any particular attribute but being good overall.

Since the publication of Goldberg's book [Goldberg 1989] in which he defines a method called non-dominated sorting, the direction of research has changed. Techniques have started to use the concept of Pareto optimality through selection and ranking methods.

The selection methods use the concept of Pareto optimality to select individuals via an elitist or tournament selection procedure. As such procedures give preferential treatment to the fittest members of a population, in the elitist method Pareto individuals are passed directly into the next generation or used in crossover and mating. Using tournament selection Pareto competitions are held between parents and their offspring and the two winning strings propagate to the next generation. Gero applied this type of technique to structural optimization [Gero 1995].

Ranking methods are used to grade the population in terms of Pareto dominance. Goldberg suggested non-dominated sorting [Goldberg 1989]. This method involves first finding all of the Pareto optimal points within a population, giving them a rank of one and removing them. The remaining population members are again processed to find non dominated individuals and these are given rank two and removed and so on until all of the population has been ranked. Both Richardson (Richardson 1990)] and Ritzel [Ritzel 1994] implement this type of ranking. Fonseca and Fleming [Fonseca 1993] suggest another ranking scheme called multiobjective ranking. In this scheme each individual is given a rank according to how many individuals dominate it. If an individual is non-dominated it is given a rank of 1, but if five individuals dominate it then a rank of 6 is assigned and so on. This method gives a greater range of ranks than Goldberg's approach and will also penalise areas of high density solutions.

(c) Combined Search and Multiple Criteria Decision Making

In this approach search and multiple criteria decision making are combined to form an iterative solution approach. The basic pattern is as follows:

(i) Perform multiple criteria search to obtain an approximate idea of the Pareto surface.

(ii) Apply multiple criteria choice or ranking to capture the preferences of the decision maker. Return to (i) but let the search be informed by the information on the priorities of the DM.

This process is continued until the final solution is selected. Fonseca and Fleming adopt a goal attainment method in their multi-objective GA (MOGA) that uses a "distance to target" approach to combine search and decision making. The target is chosen by the decision maker after examining the initial collection of solutions from a preliminary search. More recently the approach of the FRONTIER [Frontier 1996] project has been to combine multi-objective genetic algorithms with additive utility functions to form an interactive solution strategy. In this approach the decision maker is asked to compare pairwise a collection of non-dominated solutions obtained from a preliminary search using a multi-objective GA. The information on these pairwise comparisons result in one of two kinds of ordering:

- Solution i is preferred to Solution j
- Solution i is considered to be just as attractive as Solution j

The first ordering is a "preference" relation and the second an "indifference" relation. Using these relations a set of step-wise linear marginal utility functions (i.e. utility function for each criterion in turn) can be derived analytically for all the attributes of a decision (or design). The marginal utility functions are then combined to form a composite fitness function [see Section 4.3.5]. This is then used by the multi-objective genetic algorithm to concentrate its search in the next iteration, as Figure 5.7 shows.

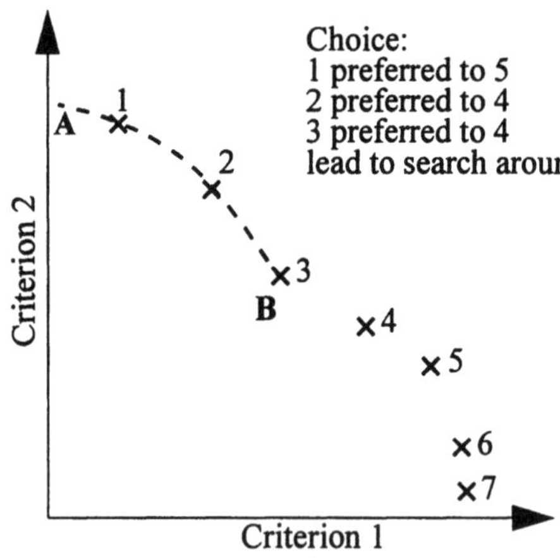

Figure 5.7 *A combined strategy of MCGA Search and multi-attribute decision making*

As the composite fitness function captures a DM's preference structure, the new search will tend to concentrate on those areas of the Pareto surface that are in harmony with his sense of priorities. As new features of interest emerge during the search, the decision maker can communicate his revised priorities by pairwise comparisons of a sub-set of the emerging solutions. This will then trigger a recomputation of marginal utility functions and a new direction of emphasis for the search.

5.3.2 Common Issues in Multiple Criteria Genetic Algorithms in Engineering Design

It is clear on the basis of the above that all three types of approaches to multiple criteria genetic algorithms essentially depend on alternative search and multiple criteria selection. As argued in some of the earlier Chapters, engineering design decision making consists either of synthesis on the basis of prioritised objectives or selection on the basis of prioritised attributes. It is clear, therefore, that there is a close harmony between the philosophy of multiple criteria genetic approaches and the fundamental concerns of engineering design, where a search is related to synthesis and a selection is exactly what it says it is.

The search process, however, is itself capable of being further subdivided into two distinct types, each with its own characteristics that require specific forms of representation and genetic operators. These types may be described as:

(i) *Parametric search problems*
This involves the standard GA as the problem is to find optimal values of variables (or parameters) to yield the best value of the fitness function. The usual approach is therefore one of binary (or some other) encoding with selection, n-point crossover and mutation to reconstruct strings.

(ii) *Combinatorial search problems*
These problems are position oriented in that the optimal ordering or sequencing of solutions is what is required. To handle this class of problems more complex encoding is required than for the conventional GA. The search works on the basis of reconstructing adjacency relationships within a string. To do this special crossover and mutation operators have to be used.

Some complex problems involve a combination of parametric and combinatorial search. For example, a factory layout may be decided on by a combinatorial approach and the work stations optimised for capacity by the parametric approach.

There are many operators that can be used but the principles are easily understood on the basis of some simple operators presented by Murata [Murata 1994].

Crossover

In the simplest one point crossover operator for combinatorial or sequencing problems, a crossover point is selected in one of the parents at random, and the genes to the left of this point are copied straight across to the child. The rest of the child is constructed using the same order as in parent 2 with the genes already picked in the first parent removed.

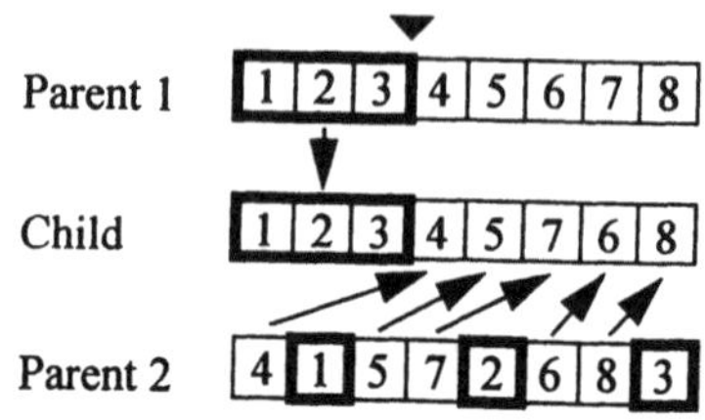

Murata's mutation operator is just as simple. In this case a gene is selected at random in the parent and moved to a random location in the child. The rest of the genes are fitted by shifting along to the right.

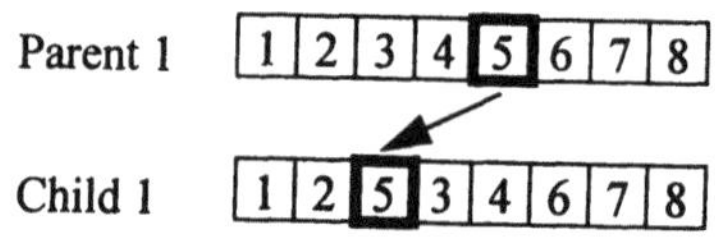

It is clear that both operators use the same number only once in the children produced, thus maintaining the requirement of non-repeatability.

5.3.3 Crowding and Niching

One of the aims of multiple criteria optimisation is to generate and maintain a diverse range of Pareto optimal solutions. In order to promote this diversity a technique originally used in single objective GAs can be employed. This is the method of Fitness Sharing or Niching, originally used by Goldberg to find multiple regions of efficiency over a multi-modal search space. The technique employed encourages individuals in the population to move away from crowded areas.

It does so essentially by finding the number of individuals in a given individual's niche and if this is n then fitness of the given individual is given a new fitness value of 1/n times its original value. It is obviously necessary to declare the niche size before the search begins.

Niching has been used in several multiple criteria approaches. The Multiple Objective Genetic Algorithm (MOGA) of Fonseca and Fleming uses multiobjecive ranking and performs fitness sharing between sets of solutions of the same rank [Fonseca 1993]. The Niched Pareto Genetic Algorithm (NPGA) holds a type of binary tournament selection called a Pareto domination tournament [Horn 1993]. In this tournament two strings are compared to a sample of the population and the number of dominating points in the sample is counted for both individuals. The individual with the least number of dominating points survives. If both points have the same number then the one with the lower niche count survives. The size of the sample used can be varied depending on the level of selection pressure required. Obviously larger sample sizes increase the bias towards stronger solutions. This method is thus analogous to performing a local version of the ranking method of Fonseca and Fleming. Both MOGA and NPGA perform sharing within criteria space. A third method, the Non-dominated Sorting Genetic Algorithm (NSGA)[Srinivas 1995] uses Goldberg's ranking method. However, sharing is performed in the phenotypic space by measuring the vector distance between the decoded design variables.

Niching can and is also used for enforcing mating restrictions between dissimilar individuals in the population. This is to prevent low fitness individuals arising out of the combination of features from high fitness individuals of vastly different properties. As in fitness sharing a niche size is defined within which an individual is allowed to mate. If a suitable partner of sufficient fitness cannot be found within the niche a partner may then be selected at random.

5.3.4 Estimating Niche Sizes

It is obviously necessary to agree on the niche size before the GA search commences. As it is beneficial in multiple criteria decision making terms to adequately cover the Pareto surface by sampled solutions it is desirable to have an even spread of points over this surface.

An intuitive approach to niche sizing can be conveniently examined by considering a bi-criteria problem. As Figure 5.8 shows the number of points required to define a Pareto surface depends on three factors:

(i) the shape of the Pareto surface
(ii) the niche size, represented by σ_{share}
(iii) the number of criteria

The shape of the Pareto surface depends in turn on the minimum and maximum values of the criteria in question (m_1, m_2 and M_1, M_2 respectively for the two criteria) and the distance of the Pareto surface from the origin (m_1, m_2). As the Pareto surface moves out from the diagonal $0_1, 0_3$ towards $0_1, 0_2, 0_3$ the number of points required to represent the surface obviously increases. The influence of niche size on the number of points required is simple to relate to. As σ_{share} reduces in value the resolution increases and the number of points required also increases.

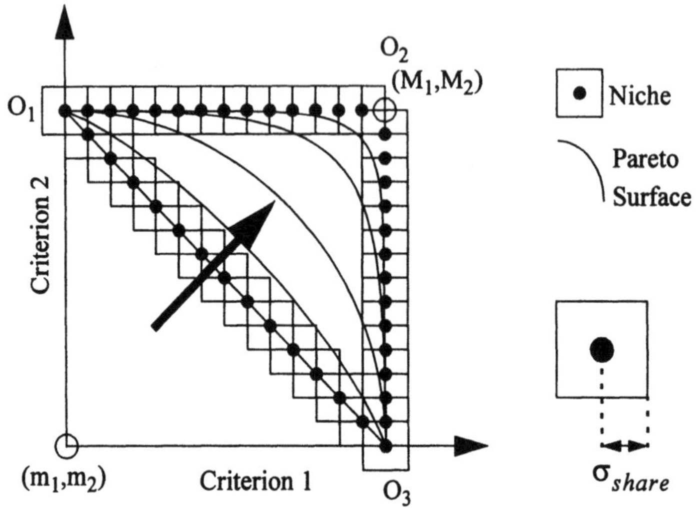

Figure 5.8 *Defining a bi-criterion Pareto Surface*

If σ_{ishare} is defined as the niche size for criterion *i*

$$\sigma_{ishare} = \frac{M_i - m_i}{r}$$

where *r* is the resolution required. As Figure 5.9 shows, r is simply the number of points required to represent the surface in criterion direction. Thus, if N is the maximum number of points required to specify the Pareto surface $N = ir^{i-1}$.

This provides a useful limit to the number of individuals in the population.

On implementing the above algorithm on some problems it becomes clear that another observation can be made. This is that in some problems many rank 1 solutions are lost due to genetic operations on them. For such problems it is sensible to maintain an off-line population of rank 1 solutions in addition to the normal population. This second population of currently rank 1 solutions may be called the Offline Population.

For every generation rank 1 solutions are copied to the Offline Population. This population is then genotypically checked to see whether any duplicates exist. Genotypic comparison is used as more than one string or individuals may have the same combination of criteria values and it would not be sensible to eliminate all such solutions. The Offline Population can then be checked to see if there are any dominated solutions as may be expected because the GA produces fitter solutions with the passing of the generations.

Number of criteria:

2	3	4	i

Number of points to reprsent a Pareto surface:

$2r$	$3r^2$	$4r^3$	ir^{i-1}

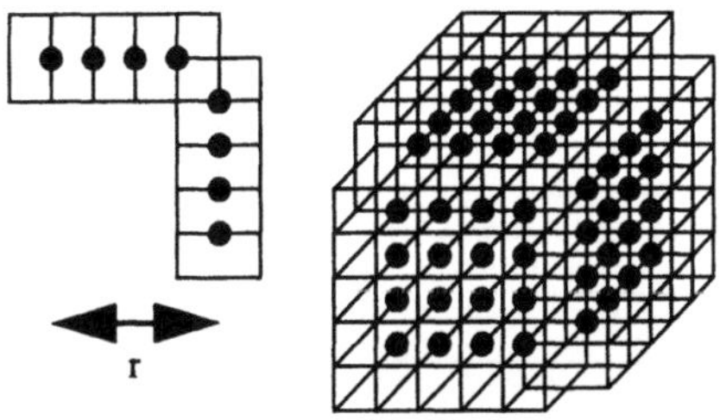

... ...

Therefore number of points required (resolution, r = 4):

8	48	256	$i4^{i-1}$

Figure 5.9 *Defining a general Pareto surface*

This Offline Population can be used for a variety of purposes. It can be used, for example, to compute the rate of generation of new efficient solutions and use that as a stopping criterion. Again, as the Offline Population is a store of the current individuals on the Pareto front, maximum and minimum values of each criterion can be chosen from this population for computing niche sizes as described above.

5.4 The Multiple Criteria Genetic Algorithm (MCGA) A Summary

It is clear on the basis of the above discussions that the Multiple Criteria GA approach described here incorporates some key features. These may be summarised as in Figure 5.10.

The MCGA method can then be summarised as a combination of the following steps

1. Create population with associated string representation:

This population is randomly generated within the feasible region. The string representation is to allow the usual operators to function as intended.

2. Evaluate population on all criteria:

These are naturally problem related as they define the various criteria.

3. Rank population members using dominance:

Using a ranking scheme like that of Fonseca and Fleming described in Section 5.3.1 the MCGA deals with several criteria simultaneously. It is important to note

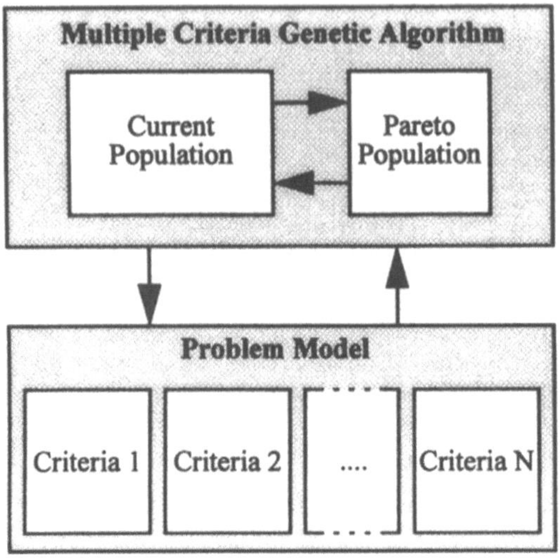

1. Create population with associated string representation.
2. Evaluate population on all criteria.
3. Rank population using dominance.
4. Update the Pareto Population.
5. Perform Fitness Conversion.
6. Perform Fitness Sharing.
7. Selection:
    Step 1: Elitist strategy.
    Step 2: Roulette Wheel Strategy.
8. Perform Restricted Crossover.
9. Perform Mutation.
10. Evaluate population on all criteria.
11. If not converged go to 3..
12. Output Results

Figure 5.10 *The Multiple Criteria Genetic Algorithm*

that the concept of dominance (defined as superiority with respect to at least one criterion without simultaneous inferiority in any other criterion) allows simultaneous consideration of the range of criteria without combining them explicitly.

4. *Update the Offline Population:*

The Offline Population maintains a full set of current rank 1 (or non-dominated) solutions. This population is updated at every generation with new non-dominated solutions and any duplicates or dominated solutions are eliminated.

This population can be used for niche sizing which indirectly controls fitness sharing and mating.

5. *Perform fitness conversion:*

The rank of individuals in a population can vary between 1 (non-dominated) to population size (totally dominated). This information can be converted into fitness functions of various types. One possible fitness function could be

$$\text{F} \equiv \text{Fitness} = ((PS + 1 - R_j)^a$$

Where PS = population size

R_j = rank of individual j

a = control factor to adjust selection pressure (may be set to 1 initially)

6. *Perform fitness sharing:*

If m_1 and M_1 can be computed from the initial population and r agreed upon, fitness sharing and mating restrictions may be applied by counting the number of individuals in a particular individual's niche.

As Figure 5.11 shows above, the number of individuals in a chosen string's niche is counted first. For the string in question this count is equal to 5, so the fitness for this string can be set at

$$F_{new} = F / 5$$

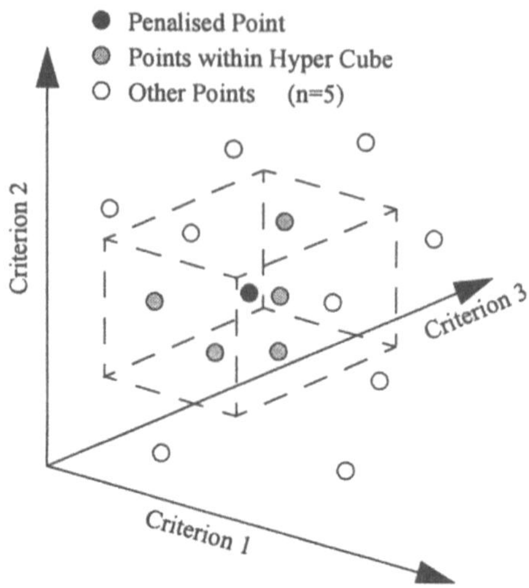

Figure 5.11 *Fitness Sharing*

7. *Selection:*

<u>Step 1</u> The Pareto efficient individuals within the population are passed directly through to the mating pool. This is the elitish strategy flagged in Figure 5.10.

<u>Step 2</u> The remainder of the mating pool is filled with members of the current population using a fitness biased procedure like a roulette wheel selection.

8. *Perform restricted crossover:*

Using the current niche size, individuals are restricted to mating with others in the same niche as themselves.

9. *Perform mutation:*

This is self-explanatory. Mutation is only comparatively rarely performed when compared to crossover.

10. *Evaluate population on all criteria:*

Same as Step 2 above, except evaluation is carried out for new population.

11. *If convergence go to 12 otherwise go to 3:*

If convergence criteria are met then output results. Otherwise repeat from Step 3.

5.5 A Numerical Example

The implementation scheme discussed above summarises the principal issues related to the practical use of the MCGA. It is clear from the preceding discussion that the implementation of any GA, by virtue of its essentially intuitive approach, is partly a matter of applying some fundamental principles and supporting this with certain common sense structures (e.g. niching, fitness sharing). Using a simple numerical example some of these issues may be illustrated.

The numerical example used is based on a hill function which is overlaid several times on each other to simulate a multiple criteria decision making space, as Figure 5.12 shows.

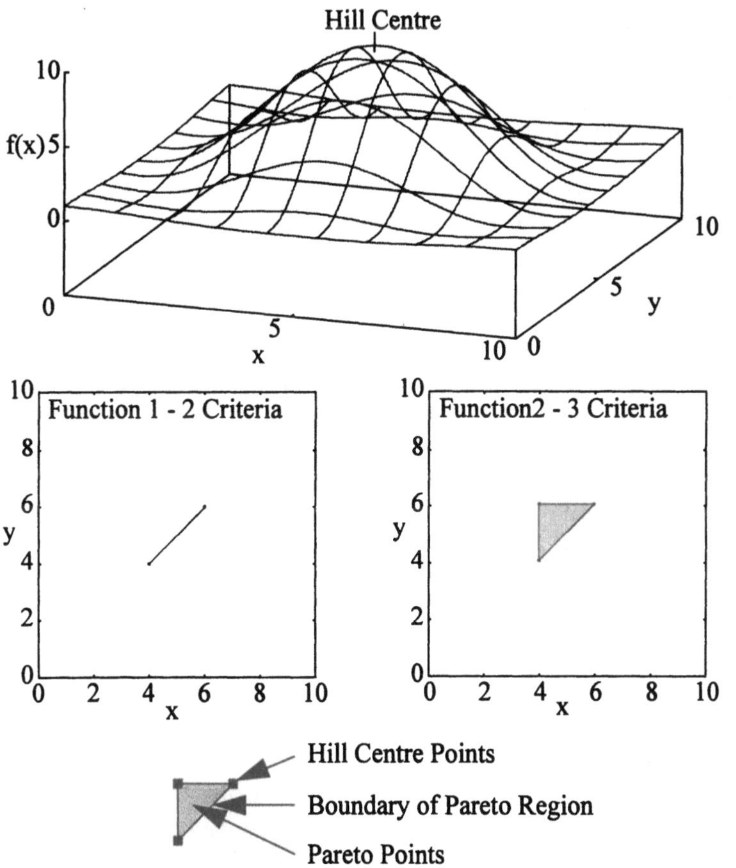

Figure 5.12 *Variable space plots of the test functions for MCGA*

The basic evaluation function is represented in the variable space (x, y) as

$$F(x, y) = \cos(((x-5)^2 + (y-5)^2)/10)*1/\exp(((x-5)^2 + (y-5)^2)/10)*8+1$$

Figure 5.12 demonstrates how treating each hilltop as the maximum value of a different criterion a multiple criteria problem can be created, where the line joining the two hilltops in the bi-criteria problem and the shaded region for the three-criteria problem represents the Pareto solutions. These two problems are defined as functions 1 and 2 respectively.

The MCGAs performance capability may be tested by examining how the population converges to the Pareto surface, using Test Functions 1 and 2. For this purpose the following parameters are kept fixed.

Population size = 500
Number of generations = 100
Crossover probability = 0.7
Mutation probability = 0.01

Thus the total number of evaluations is 500 x 100 = 50,000. As each of the two variables is represented by a 10-bit number, the total search space consists of 1024^2 solutions. The GA, therefore, only evaluates 4.8% of the possible combinations. Resolution r was determined for each problem for N = 500 and using i = 2 and 3 respectively.

Figures 5.13 and 5.14 show the final location of the members of Offline Population and their distribution over the x-y variable space. The plots clearly demonstrate that the MCGA is capable of finding the Pareto regions quite effectively. In the ideal situation the Pareto solutions would be uniformly distributed over the region of interest and all peaks would be equally high. However, this is not always the case as some spaces are more intensively sampled than others. This is partly the result of the choice of initial population and partly a consequence of the search strategy in the form of crossover, mutation and niching restrictions.

These influences are shown in Figure 5.15 for the normal population for the bi-criteria problem. The initial population is distributed widely and there are very few individuals near the Pareto surface.

By Generation 10 clustering appears on the Pareto front, although there are still some gaps on this front. In other words the population has replaced less fit individuals with fitter ones. This results in peaks along the Pareto front as the circles on Figure 5.15 shows.

By Generation 50, virtually all the solutions are on the Pareto front. In other words genetic fitness and diversity has been simultaneously propagated by the MCGA.

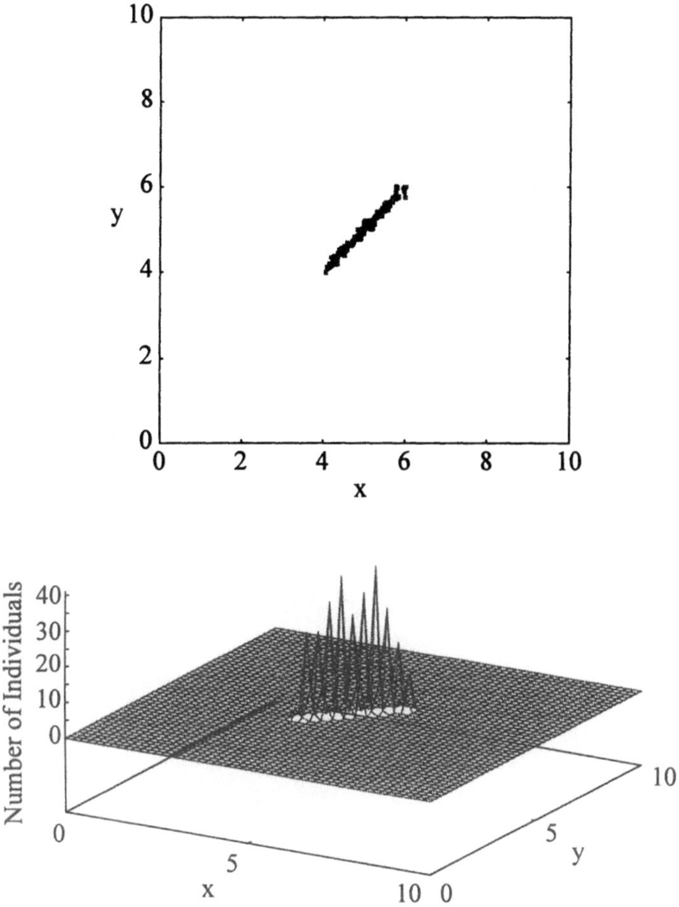

Figure 5.13 *Test Function 1 X-Y and Distribution Graphs (Generation 100)*

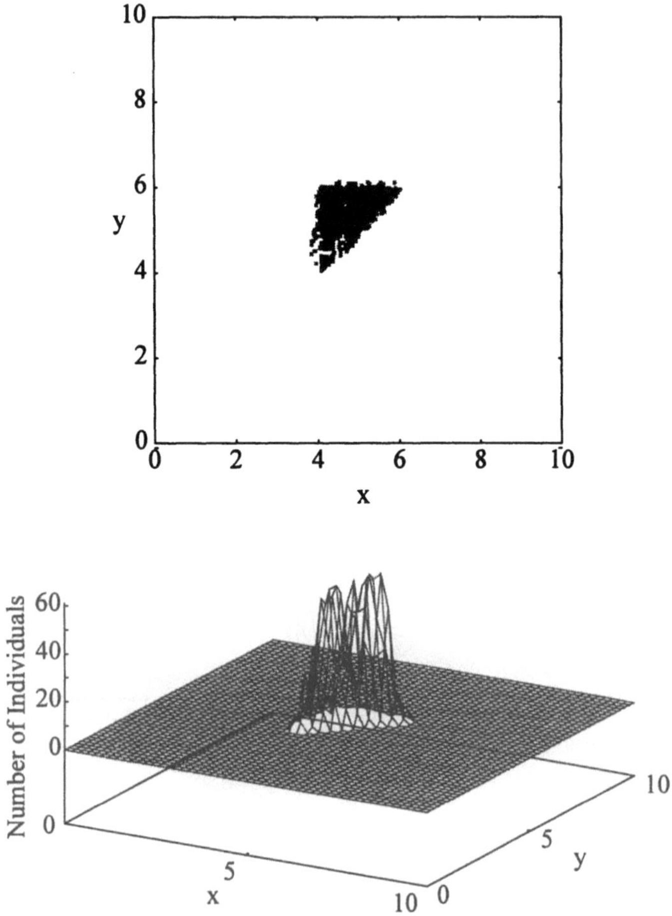

Figure 5.14 *Test Function 2 X-Y and Distribution Graphs (Generation 100)*

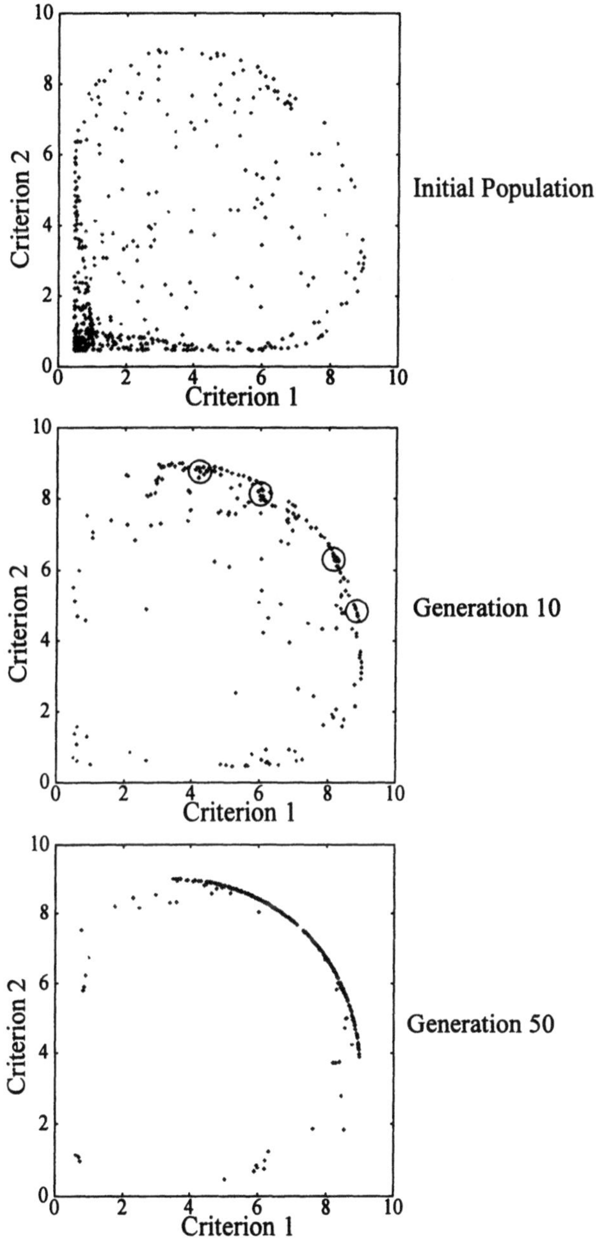

Figure 5.15 *Development of Normal Population in Criteria Space for Test Function 1*

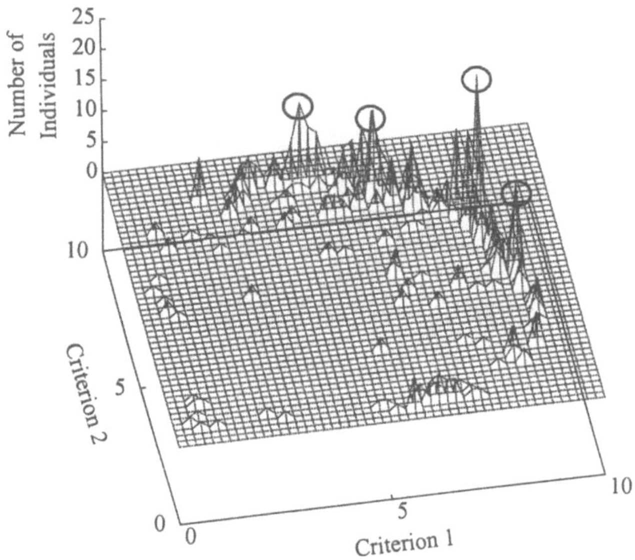

Figure 5.16 *Test Function 1 - Distribution in criteria space (10 Generations)*

Figure 5.16, shows a landscaped view of how the normal population looks at Generation 10 for the bi-criteria problem. This is in effect a three dimensional view of the second plot in Figure 5.15. The emerging Pareto front is clearly visible as the outermost range of peaks. It is discernible that there are clusters of peaks in some regions, shown circled, which is largely the result of restricted crossover. There are, however, still a fair number of solutions that are not on the Pareto front. These two characteristics are naturally addressed by the combined influence of cross-over and adaptive niche sizing.

5.6 An MCGA Schedule for a Generalised Job Shop

Scheduling is an important function in most large scale manufacturing environments. It is desirable to schedule activities in such contexts as it helps to reduce lead times, recognise key resource constraints and generally to improve manufacturing efficiency. As scheduling involves management of the myriad inter-connections between various activities it is not surprising that this has been an area of work that has received a lot of attention. Much of the attention has attempted to either simplify the problem so that formal optimisation methods like Linear Programming is applicable or used heuristics or Monte Carlo Simulation to deal intuitively with the task in hand.

All scheduling is essentially combinatorial in nature, so it is obviously worth attempting to deal with such problems using a Multiple Criteria Genetic Algorithm approach.

Before attempting to do so it would be instructive to examine the principal features of a generalised job shop scheduling problem so that the problem formulation and associated data requirement can be addressed. These features may be listed as:

(i) There are a number of machines (or workstations) and a number of different jobs.
(ii) Each job consists of a number of stages.
(iii) Each stage of each job is characterised by a specific operation.
(iv) Each machine may, in principle, be able to perform more than one operation.
(v) Each operation has a standard processing time, which is different for the different machines.
(vi) The stages of any job must be carried out in a specified order.
(vii) A job can visit a machine more than once.
(viii) Operations can only be interrupted at specified times.
(ix) A machine can only handle one job at a time.
(x) Basic processing times and due dates are known.
(xi) Machine maintenance times and set-up times are known.
(xii) Due dates must be taken into account and penalties may need to be considered.

In addition there may be more than one criteria of performance to consider, making generalised scheduling a large combinatorial multiple objective problem.

To see how such large scale combinatorial problems can be handled using MCGA, an example is outlined below.

5.6.1 Problem Data

Table 5.2 shows the types of machines and operations and the machine operating efficiency values. In this example there are four machines or workstations (M0, M1, M2, M3) and six types of operations ($OP_0, OP_1, OP_2, OP_3, OP_4, OP_5$). An efficiency of 1 indicates standard processing time with a value less than 1 implying less efficient operation. Thus, OP1 is best performed by machine 2 (100% efficiency) with machine 1 being half as efficient (50% efficiency) and hence it takes twice the time to perform the same job.

Table 5.2 - *Machine Operating Efficiency Table*

Machine	Op0	Op1	Op2	Op3	Op4	Op5
M0	1.0	0	0	0.5	0	0
M1	0	0.5	0	0	0.8	0
M2	0	1.0	0.8	0	0	0.8
M3	0.5	0.8	0	0.3	0	1.0

Table 5.3 - *Machine Maintenance Times*

Machine	Num	N/A 0	N/A 1	N/A 2	N/A 3
M0	*2*	50-60	100-110		
M1	*1*	20-25			
M2	*4*	10-15	25-30	34-37	50-52
M3	*2*	20-35	60-64		

Machine maintenance times are shown in Table 5.3. The variable Num indicates how many periods of non-availability exists within the overall period of interest. Thus machine 1 has one anticipated period of non-availability and this is between periods 20 and 25 (i.e. 5 time units).

Table 5.4 – *Operation Setup Times*

	Op0	Op1	Op2	Op3	Op4	Op5
Op0	0	4	2	2	3	1
Op1	3	0	1	5	4	2
Op2	3	2	0	2	3	1
Op3	4	3	2	0	2	1
Op4	3	5	2	1	0	2
Op5	5	4	4	1	2	0

The set up times required to switch a machine from one operation to another is shown in Table 5.4. It is clear that such a table has 0 along the diagonal as this implies continuing with the same activity (i.e. no switching of operations).

A job is defined as having a number of stages associated with it. This is shown in Table 5.5.

Table 5.5 - *Job-Stage Table*

Job	Num	Stage 0	Stage 1	Stage 2	Stage 3	Stage 4
Job 0	2	2,4	4,10			
Job 1	4	1,6	2,3	5,8	0,3	
Job 2	3	4,5	1,3	2,8		
Job 3	4	0,7	2,7	4,3	5,1	
Job 4	5	1,4	0,5	1,7	3,9	4,6
....	....	....	....	....	....	....

Each cell in the table consists of two entries; the first is the type of operation and the second is the standard time required to perform the operation. The data in Tables 5.2 and 5.5 may be combined to estimate the duration of individual operations on the different machines.

Thus operation 0 can be performed by machines 0 and 3 according to Table 5.2. However, machine 3 is half as efficient as machine 0. As Table 5.5 shows that Job 3 Stage 0 is operation type 0 then the time required to do task on this machine type 3 will be given by

$$\frac{\text{Actual}}{\text{time}} = \frac{\text{Standard time}}{\text{efficiency}} = \frac{7}{0.5} = 14 \text{ time units}$$

Finally Table 5.6 shows specific start times, dependencies, deadlines (or due dates) and associated penalties for a delay of 1 time unit.

Table 5.6 - *Job Data*

Job	Start	Num	Dep 0	Dep 1	Dep 2	Deadline	Penalty
Job 0	10	3	6	8	9	100	1
Job 1	0	0				50	1
Job 2	5	0				90	1
Job 3	0	2	7	8		100	5
Job 4	0	1	16			0	0
....	....	....	....	....	....	....	....

Thus Job 0 can start no earlier than Time 10 after start of schedule. It must wait for jobs 6, 8 and 9 to be complete before it can start and it should be completed by Time = 100. It attracts a penalty of 1 for each delay of 1 time unit thereafter.

5.6.2 String Configuration

Due to the special nature of the problem a structured GA is used [Dasgupta 1992]. The manipulation of the string in structured GA is similar to that of an ordinary

GA. The principal difference is when fitness is evaluated, where instead of using the whole string to determine fitness, only those parts of the string which are activated by genes at the higher level in the hierarchy are used. The rest of the string is carried along but play no part in the fitness computation. This is to avoid double counting of the same job. Thus, wherever there is a choice of machine for a certain stage of a given job and it is allocated to one of the machines, then the structured GA approach makes sure that the same task is not allocated to any other machine that can also do the task. This is analogous to the following construct.

Structure

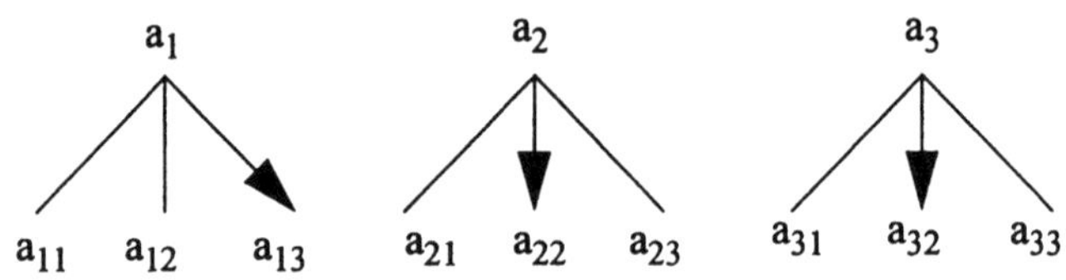

String Encoding

a_1 a_2 a_3 a_{11} a_{12} a_{13} a_{21} a_{22} a_{23} a_{31} a_{32} a_{33}

Decoded String (arrows indicate activation)

a_{13} a_{22} a_{32}

Figure 5.17 *Gene Activation*

To examine how this works in practice let us examine a small example defined by Tables 5.7 and 5.8 below

Table 5.7 – *Machine Operating Efficiency Table*

Machine	Op0	Op1	Op2	Op3	Op4	Op5
M0	1.0	0	0	1.0	1.0	0
M1	0	1.0	0	1.0	0	1.0
M2	0	0	1.0	0	1.0	1.0

Table 5.8 – *Job-Stage Table*

Job	Num	Stage 0	Stage 1	Stage 2	Stage 3	Stage 4
Job 0	3	1,5	4,3	1,4		
Job 1	5	2,3	1,6	3,5	1,3	4,4
Job 2	3	0,4	2,3	1,6		
Job 3	4	5,4	2,5	1,3	0,5	

Table 5.7 clearly shows that operations 0, 1 and 2 can only be performed on Machines 0, 1 and 2 respectively. In other words, for these operations there is no choice of machine. Operation 3 can be performed on machines 0 and 1, Operation 4 can be performed on machines 0 and 2 and Operation 5 can be performed on machines 1 and 2, all with equal efficiency of 100%. It is now necessary to identify which operations for which jobs have a choice of machines and which can be performed on one machine only. This is simple to identify as Table 5.7 shows that only if operations 3, 4 and 5 are involved in any job will there be a choice of machine.

Looking at Table 5.8 again it is clear that Job 0 Stage 1 (j0s1), Job 1 Stages 2 and 4 (j1s2, j1s4) and Job 3 Stage 0 (j3s0) involve operations 3 or 4 or 5. As j0s1 is operation type 4 and this can be performed on machines 0 and 2 [see Table 5.7], j0s1 appears on the list of machines 0 and 2 below.

Thus for the example in question, the permissible job stage allocations are:

Machine Choice according to: j0s1, j1s2, j1s4, j3s0
Jobs on Machine 0 could be: j0s1, j1s2, j1s4, j2s0, j3s3
Jobs on Machine 1 could be: j0s0, j0s2, j1s1, j1s2, j1s3, j2s2, j3s0, j3s2
Jobs on Machine 2 could be: j0s1, j1s0, j1s4, j2s1, j3s0, j3s1

It is now necessary to translate the above into a string that can be accommodated and manipulated by the MCGA. This is done by treating the machine choice section and the jobs on machines separately in the string.

For example, the four machine choices are: j0s1, j1s2, j1s4, j3s0.

Taking the first choice, j0s1 is operation type 4. So this task can only be performed by machine 0 or 2 as Table 5.7 shows. Therefore possible values of j0s1 are 0 and 2. Similarly j1s2 is operation type 3 and this operation can be performed on machines 0 and 1. Completing the whole list of four choices it is clear that the following range of values is permissible.

Choice:	j0s1	j1s2	j1s4	j3s0
Operation type:	4	3	4	5
Range of Machines of choice:	[0,2]	[0,1]	[0,2]	[1,2]

Thus separating a string into its "machine choice" and "jobs on machine" sections one sample string that could arise from the above could be

Machine Choice	Machine 0	Machine 1	Machine 2
	31210	21103301	310132

As the list of machine 0 contains j0s1, j1s2, j1s4, j2s0, j3s3 the jobs list for this machine is

Job 0	1 Sample
Job 1	2 Samples
Job 2	1 Sample
Job 3	1 Sample

Thus, any permutation of 01123 will be a legal contribution to the string. Similar arguments hold for the other machines. However, this allocation of jobs will include duplicates as the consequence of machine choice has been thus far to allocate the same tasks to all the machines capable of doing them.

Using the machine choice section of the string the jobs in each of the machine lists can be activated as follows:

The first gene in the machine choice list is for Job 0 Stage 1. This can go to machine 0 or 2 as seen above. In this case machine 0 has been chosen. It must thus be removed from the list of machine 2. Machine 2 currently has only one instance of Job 0 and this is removed. If there is more than one stage of the same job on a certain machine then implicit job ordering has to be used.

For example, the second gene on the machine choice list is Job 1 Stage 2, capable of being performed on machines 0 or 1. As 0 has been used in this string, this task must be removed from the list of machine 1. Machine 1 has 3 Job 1 genes. As machine 1 can perform Job 1 Stage 1 and Job 1 Stage 2, the Second 1 is removed from the machine 1 list.

Proceeding in this manner the following preference list is obtained for the machines

Machine 0 preference list: 3 1 2 1 0
Machine 1 preference list: 2 1 0 3 0 1
Machine 2 preference list: 3 1 3 2

Thus, this list is now free of duplicates and can be used to build an associated schedule. The list can also be used to perform crossover and mutation.

5.6.3 The Results from MCGA

The string obtained above is converted into a schedule using some logical rules. The schedule building exercise consists of starting at Time = 0 and going through until all jobs are complete. For every time increment it checks for the current state of the set-up. Using the preference list of each machine as obtained above it checks whether any machine is idle. If so it finds the highest job on the list which is waiting and executing this. A job is considered to be available for execution if it meets all of the following conditions:

(i) all previous stages of that job are complete
(ii) all dependencies are satisfied

(iii) the machine has been set-up for the job stage (i.e. operation type)

When all the jobs are complete the loop is complete and various statistics are computed. These relate to makespan (the total elapsed time to complete all jobs), average job times (time required on average to complete a job), delay times, penalty costs, machine utilisation and other related metrics.

For the purposes of this application two criteria are used : makespan and average job times. The result of running the MCGA is as shown in Figure 5.18, using a population of 500 and the concept of dominance.

The development of the Pareto front is clear, along with the nature of the trade-off between the criteria chosen. To reduce makespan the scheduler tries to move jobs around to make the best use of the machines. This leads to an increase in the average job time. To reduce the latter schedules need to group several stages together so that they can be more quickly performed in sequence. This will lead to significant gaps in the schedule where jobs are waiting until all resources are available for uninterrupted runs to take place for each job, leading to poor makespan values.

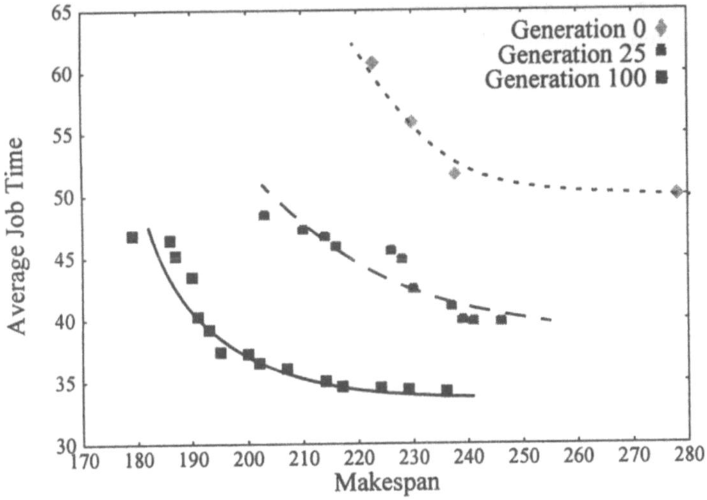

Figure 5.18 *Pareto front generation for scheduling example*

5.7 Concluding Comments

It is clear from the above that the use of adaptive optimisation methods like genetic algorithms in the context of multiple criteria decision making is a natural extension of more established methods. The method is effective not only for demanding parametric optimisation problems but also for large combinatorial ones. As parametric optimisation problems are converted into combinatorial ones by using binary or other similar representation it is enough to say that multiple criteria evolutionary search is an effective way of determining the Pareto surface, particularly for large, multi-peaked, discontinuous criteria landscapes.

In many real life problems additional benefits may be obtained from hybrid approaches that combine evolutionary search techniques with other methods. For example, having searched widely using GAs' some hill-climbing algorithms may be used in the vicinity of the "best" solution so far to determine peaks of interest. Again, multiple criteria decision making over a widely distributed subset of solutions obtained using GAs can be used to capture the DM's underlying preference structure. This may be done by estimating marginal utility functions, for example, for the attributes. These marginal utility functions may be combined to form a composite fitness function to direct subsequent GA searches to a part of the Pareto surface, thus making the search more directed and efficient, if this is what is required.

6
An Integrated Multiple Criteria Decision Support System

6.1 System Structure and Method Selection

6.1.1 General Structure of IMC-DSS

In the previous chapters, we have discussed various MCDM techniques. Each of those techniques may be applied to deal with a different design decision problem on an individual basis. It is, however, of significant advantage that those techniques be incorporated into a software system in an integrated manner, so that the decision maker could use such a system to investigate his decision problems using several methods without being bothered by the mathematics involved.

In this chapter, the system framework and application examples of an integrated multiple criteria decision support system (simply referred to as IMC-DSS) are discussed. The framework was developed at the Newcastle Engineering Design Centre [Yang 1992a, 1992b and 1992c], [Sen and Yang 1993a and 1995a] and is being constantly added to. In the development of the system, attention is being paid to generality, portability and flexibility. This leads to the incorporation of several MCDM techniques with various features into the system. The reason for employing a range of techniques in the system is that no single MCDM method could be regarded as absolutely superior to others in every decision situation or by every decision maker. All MCDM methods may have their own desirable features as well as weak points. The method used depends to a large extent on the decision situation and the persons using it.

This IMC-DSS is composed of three main parts, as shown in Figure 6.1. The first part consists of a routine base and its management system for typical MCDM methods and some supporting techniques, which forms the kernel of the IMC-DSS and which decides what the system can do in a given situation. The second part includes a data base, a model base and their management systems. This part consists of a number of files in which raw data, subroutines for listing model functions, and intermediate and/or final results are stored. For a particular application, a semi-automatic model generator may also be included for supporting the construction of a class of decision models. Such a model generator may be established as a knowledge-based system. The third part consists of a user interface which is designed to assist in the selection of MCDM methods, data preparation and model building, interactive decision making and result analysis.

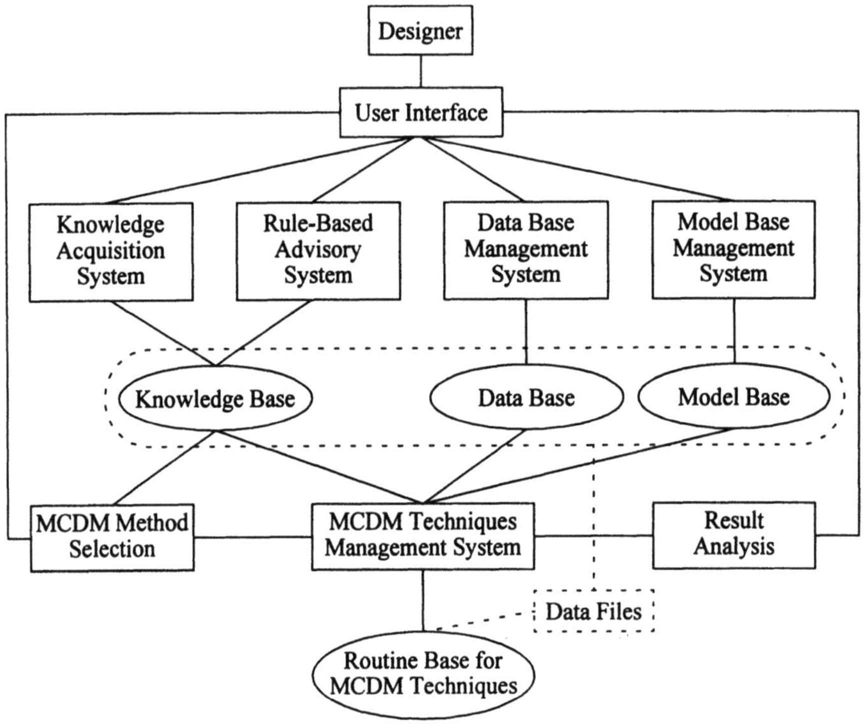

Figure 6.1 *System Framework of the IMC–DSS*

6.1.2 The Routine Base for MCDM Techniques

In the routine base of the IMC-DSS, eight MCDM techniques are organized together in an integrated manner, including four MADM and four MODM methods. The four MADM methods are TOPSIS, CODASID, AHP and UTA methods, and the four MODM methods are minimax (ideal point), ISTM, goal programming and Geoffrion's methods. These methods have been described in the previous chapters. In addition, there is a separate environment for running the multiple criteria genetic algorithm (MCGA) described in Chapter 5 [Todd 1997].

The routine base is a computer software system composed of a large number of programs including both C functions and FORTRAN subroutines which are stored in a number of files. Each of the MCDM methods was initially coded separately and could be run on an individual basis. However, input data structures for these individual programs are so designed that the programs can share similar data structures. For instance, the programs for the MODM methods share the same input data structure. Such a design strategy allows for an easy extension of the system in line with current and future developments.

The internal structure of the routine base is illustrated by Figure 6.2. As MODM and MADM problems are two distint classes of decision problems, they are treated in different ways in the IMC-DSS. A MADM problem represented by means of a decision matrix can be solved using either the TOPSIS method or the CODASID method or the UTA method. A MADM problem represented by pairwise comparison matrices can be solved using the AHP method. Solutions obtained using these four methods are the rankings of candidate designs. The C programs for each of these methods were initially developed separately and then assembled together to construct a decision support software sub-system for multiple attribute decision making [Yang 1992a]. In the subsystem, the three methods, TOPSIS, CODASID and UTA, share the same input data structure.

In the IMC-DSS a MODM problem is represented as a vector (multiple objective) mathematical programming problem as shown in (4.1). The four MODM methods, the minimax, ISTM, goal programming and Geoffrion's methods, are used to transform a multiple objective problem into a sequence of single-objective problems, whose optimal solutions are the efficient solutions of the original multiple objective problem. Since the minimax and goal programming methods require *a priori* overall preference information, they only carry out the transformation once. The ISTM method and Geoffrion's method are interactive MODM techniques and they implement the transformation iteratively as the procedure proceeds. In each transformation conducted in the ISTM and Geoffrion's methods, local preference information needs to be acquired based on the efficient designs already generated.

The programs for each of the four MODM methods were initially developed separately, sharing the same input data structure for representation of a multiple objective problem and also sharing the same model structure for the transformed single-objective optimization problem(s). These programs are assembled together to establish a decision support software sub-system for multiple objective decision making [Yang 1992b]. A single objective problem is solved using a sequential linear programming (SLP) routine.

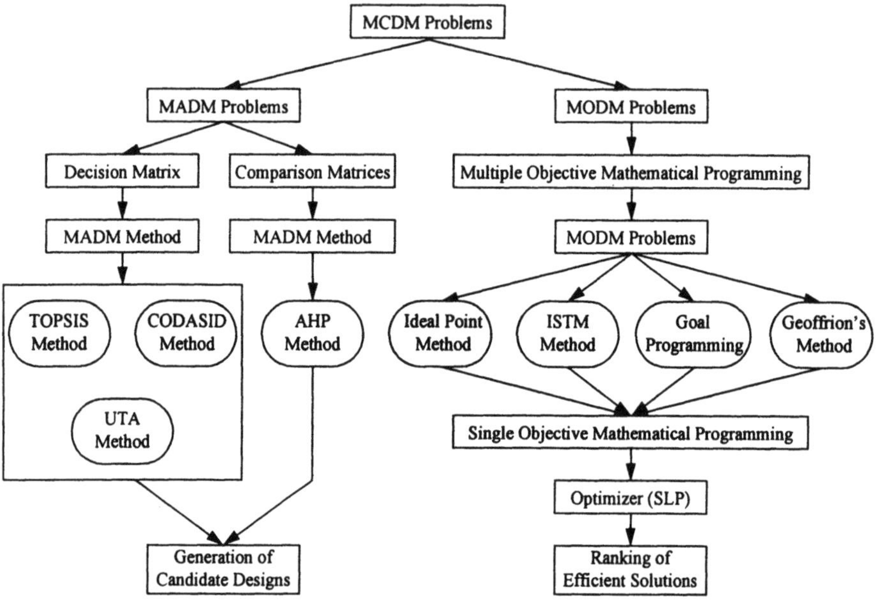

Figure 6.2 *Internal Structure of the IMC–DSS*

The above two sub-systems are then integrated together to construct the routine base for the MCDM methods.

The main advantage of the above arrangement of the internal structure for the routine base is that the routines for each of the MCDM methods can be developed separately so that adding new methods into the system will not seriously affect the programs for the existing methods.

6.1.3 Rules for Selection of MADM and MODM Methods

Each of the eight MCDM methods discussed in the previous chapters has its own features. A rule-based advisory system is designed to help the user select an appropriate MCDM method [Yang 1992c].

The selection rules for the MCDM methods are based on the general rules for classification of MADM and MODM methods, as discussed in Sections 3.1.2 and 4.1.2. The selection of the four MADM methods, as shown in Figure 6.3, is based on the following three types of questions:

1> What kind of input data is available?

2> What type of preference information can be elicited?

3> Which decision rule is adopted?

The advisory sub-system provides multiple choice answers to each of the above questions. A help option is also included for these questions and their answers. When a MADM method is chosen, the synopsis, computational steps and weak points of the method are listed. In this way, the designer is helped to better understand the method so that he may be in a proper position to judge if the chosen method is the one he really wants to use.

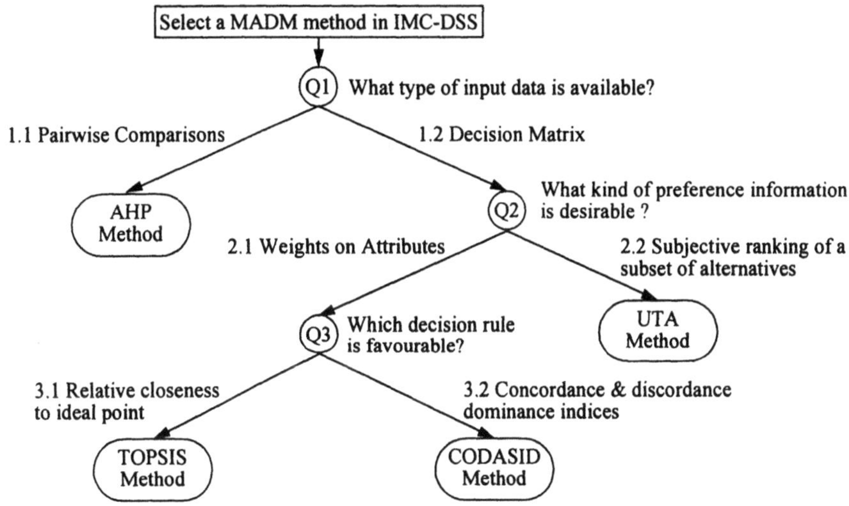

Figure 6.3 *Selection of MADM Methods in the IMC–DSS*

The selection of the four MODM methods, as shown in Figure 6.4, is based on the following three types of questions:

1> How is preference information elicited?

2> What type of preference information is available?

3> Which decision rule is adopted?

In this advisory sub-system two, multiple choice answers to each of the above questions and help options for these questions and answers are provided, along with technical details about the MODM methods.

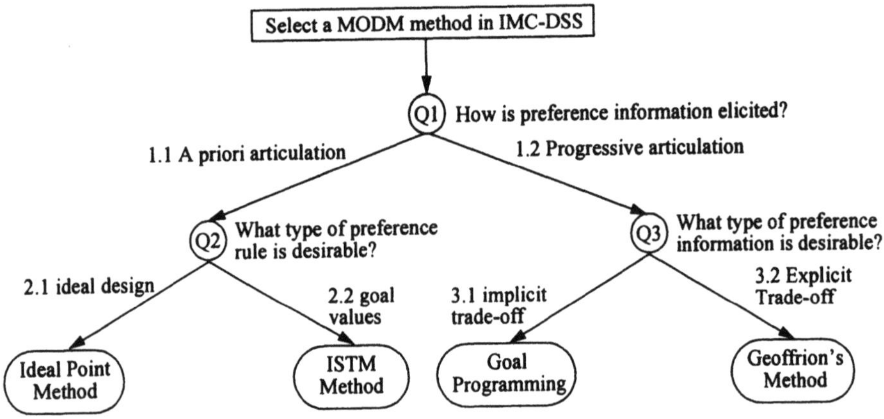

Figure 6.4 *Selection of MODM Methods in the IMC–DSS*

6.2 Data Base and Model Base

6.2.1 Decision Models and File Systems

In Figure 6.2, it is clear that three types of decision models can be handled using the IMC-DSS, that is decision matrix, pairwise comparison matrices and multiple objective mathematical programming. These decision models decide the structures of the model base and the data base of the IMC-DSS.

In a computer, decision models may be represented by a file system with specific data structures, which forms part of the data base and the model base. For instance, a decision matrix may be represented by a simple data file, listing the number of attributes, the number of alternatives and the numerical evaluation values of alternatives on attributes. For example, the aircraft selection problem, as given in Table 3.10 and equation (3.65), can be represented using a data file as shown in Figure 6.5 [Yang 1992b].

number_of_alternatives: 6

number_of_attributes: 6

2.0	1.5	2.0	-5.5	5.0	9.0
2.5	2.7	1.8	-6.5	3.0	5.0
1.8	2.0	2.1	-4.5	7.0	7.0
2.2	1.8	2.0	-5.0	5.0	5.0
2.8	3.0	2.5	-4.0	10.0	10.0
1.5	1.0	1.5	-7.0	0.0	0.0

Figure 6.5 *Data File of Decision Martrix* for *Aircraft Selection Problem*

Pairwise comparison matrices may also be represented using a data file where the information about an evaluation hierarchy may be accommodated, including the number of attribute levels, the number of attributes at each attribute level, the number of alternatives, and numerical values for pairwise comparisons between attributes or between alternatives. In the ship choice problem, as shown in Figure 3.14, for example, the pairwise comparison matrices as given in Tables 3.6-1 to 3.6-4 can be represented by a data file as shown in Figure 6.6 [Yang 1992a].

In Figure 6.6, the integer "3" in the first row denotes the third attribute level. In the second row, the integers "3" and "5" stand for the number of attributes at the third level and the number of attributes at an adjacent lower level, respectively. The first 3-dimensional matrix as given by the three rows following the integers represent the comparison matrix given by Table 3.6-1. The last three 5-dimensional matrices represent the comparison matrices given by Tables 3.6-2 to 3.6-4.

3				
3	5			
1.000000	1.000000	1.000000		
1.000000	1.000000	1.000000		
1.000000	1.000000	1.000000		
1.000000	0.500000	1.000000	0.333333	2.000000
2.000000	1.000000	0.500000	0.333333	2.000000
1.000000	2.000000	1.000000	1.000000	3.000000
3.000000	3.000000	1.000000	1.000000	2.000000
0.500000	0.500000	0.333333	0.500000	1.000000
1.000000	0.333333	2.000000	3.000000	2.000000
3.000000	1.000000	3.000000	3.000000	2.000000
0.500000	0.333333	1.000000	4.000000	2.000000
0.333333	0.333333	0.250000	1.000000	1.000000
0.500000	0.500000	0.500000	1.000000	1.000000
1.000000	3.000000	4.000000	4.000000	1.000000
0.333333	1.000000	2.000000	2.000000	0.500000
0.250000	0.500000	1.000000	1.000000	0.333333
0.250000	0.500000	1.000000	1.000000	0.333333
1.000000	2.000000	3.000000	3.000000	1.000000

Figure 6.6 *Data File of Comparison Matrices* for *Ship Choice Problem*

In the model base, a multiple objective mathematical programming problem is represented using the following general form

$$\min \quad F(X) = \{f_1(X) \cdots f_i(X) \cdots f_k(X)\}$$

$$s.t. \quad X \in \Omega \qquad X = [x_1 \cdots x_n]^T$$

$$\Omega = \left\{ X \;\middle|\; \begin{array}{l} g_i(X) \le 0,\ h_j(X) = 0,\ \ i=1, \cdots, n_1,\ \ j=1, \cdots, n_2 \\ s_l(X) \le 0,\ e_m(X) = 0,\ \ l=1, \cdots, n_3,\ \ m=1, \cdots, n_4 \end{array} \right\}$$

where x_i is a design variable, $f_i(X)$ a linear or nonlinear objective function, $g_i(X)$ a nonlinear inequality constraint function, $h_j(X)$ a nonlinear equality constraint function, $s_l(X)$ a linear inequality constraint function and $e_m(X)$ a linear equality constraint function. The separate consideration of linear and nonlinear constraints is due to the use of sequential linear programming in the routine base.

In the model base, several C functions are designed for defining objective functions and linear equality, linear inequality, nonlinear equality and nonlinear inequality constraint functions, which are named o_funcs_def(), l_eqs_def(), l_ineqs_def(), nl_eqs_def() and nl_ineqs_def(), respectively.

Suppose the following problem is identified as a model in the model base,

$$\begin{aligned}
\max\ & f_1(X) = 2.0 + 0.524(x_1-0.3) + 2.79(x_2-0.3) \\
& \qquad + 0.882(w_1 - 0.3) + 2.65(w_2 - 0.3) \\
\max\ & f_2(X) = 7.5 - 0.012(59/(1.09 - x_1^2) - 59) \\
\min\ & f_3(X) = 1.8\times10^{-3}(532/(1.09 - x_2^2) - 532) \\
s.t.\ & X \in \Omega \qquad X = [x_1 \ \ x_2 \ \ x_3]^T
\end{aligned}$$

$$\Omega = \left\{ X \left| \begin{array}{l}
g_1(X)\text{: } 4.75 + 2.27(x_1 - 0.3) \geq 6.0 \\
g_2(X)\text{: } 5.1 + 0.177(x_1 - 0.3) + 0.978(x_2 - 0.3) + 0.216(w_1 - 0.3) \\
\qquad + 0.768(w_2 - 0.3) \geq 6.0 \\
g_3(X)\text{: } 2.50\times10^{-3}(450/(1.09 - x_3^2) - 450) \leq 1.5 \\
g_4(X)\text{: } 1.0 + 0.0332(x_1 - 0.3) + 0.0186(x_2 - 0.3) + 3.34(x_3 - 0.3) \\
\qquad + 0.0204(w_1 - 0.3) + 0.78(w_2 - 0.3) + 2.62(w_3 - 0.3) \geq 3.5 \\
0.3 \leq x_i \leq 1.0;\ \ w_i = 0.39/(1.39 - x_i^2);\ \ i=1, 2, 3
\end{array} \right. \right\}$$

The following C functions may then be used to represent the model in the model base, as shown in Figures 6.7a to 6.7e [Yang 1992b].

To run the model using any of the four MODM methods in the routine base, the following data file is designed for identifying the model, for listing the structural information about the model such as the numbers of objectives and constraints, and for assigning the upper and lower bounds, initial values and initial step sizes of variables, as shown in Figure 6.8.

6.2.2 Semi-Automatic Model Generation

In the model base, the above specific data files and C functions are used as interfaces between models and the MCDM methods of the routine base. In fact,

```
/*
** The list of the objective functions
*/

double o_funcs_def(double *X, int i, int id_model)
{
  double value;

  /* Multiple objective model 15 */
  if(id_model == 15)
  {
    if(i == 1)
      value = -2.0 - 0.524*(X[1] - 0.3) - 2.79*(X[2] - 0.3)
            - 0.882*(0.39/(1.39-X[1]*X[1]) - 0.3)
            - 2.65*(0.39/(1.39-X[2]*X[2]) - 0.3);
    if(i == 2)
      value = -7.5 + 0.012*(59.0/(1.09-X[1]*X[1]) - 59.0);
    if(i == 3)
      value = 1.8*pow(10.0, -3.0)*(532.0/(1.09-X[2]*X[2]) - 532.0);
  }

  return value;
}
```

Figure 6.7a *C Function* for *the Objectives of A Model*

the C programs in the routine base treat the above data files or C functions either as input data files or as input subroutines.

In the above data files, a decision model is only represented using abstract data. Much of the information about the model, such as the definitions of the attributes and alternatives, is not properly recorded. For a large, complex decision problem, represented for example by a decision matrix, it would not be an easy task to create, maintain and update a data file in a consistent way. This is because with the increase in the dimension of a decision matrix it will become more difficult to trace the exact meaning of individual data or the relations between data. There is therefore a need to construct a more comprehensive model base. In such a model base, not only should the above data files be involved but all other information, necessary to completely represent the decision model, should also be properly recorded.

```
/*
** The list of the linear equality constraint functions
*/

double l_eqs_def(double *X, int i, int id_model)
{
  double value;

  /* Nonlinear model 15 */
  if(id_model == 15)
  {
    if(i == 1)
      value = 0.0;
  }

  return value;
}
```

Figure 6.7b *C Function* for *the linear equality constraints of A Model*

Furthermore, the knowledge of a decision maker about a decision problem may not initially be organized by the rather formal frameworks of the above decision models. In fact, one of the first obstacles facing practitioners using multicriteria decision support systems is the gap between the usual data and knowledge of the DM and the formal decision models. A semi-automatic model generator is therefore desirable, which should be able to accept the original knowledge of the DM as input data and generate a formal decision model as output by means of rule or knowledge based reasoning. In this way, the DM's task in model building could be simplified and a decision model could be created, maintained and updated in a consistent and reliable manner. However, a complete model generator may only be built for application in a specific problem domain as the use of knowledge representation techniques and reasoning schemes depends on the type of knowledge available.

In view of the above, a data base and a model base in a MCDM based decision support system may need to be implemented using an object-oriented data base management system, which should be capable of supporting knowledge-based model building and graphic interface design for modeling and decision making.

```
/*
** The list of linear inequality constraint functions
*/

double l_ineqs_def(double *X, int i, int id_model)
{
  double value;

  /* Multiple objective model 15 */
  if(id_model == 15)
  {
    if(i == 1)
      value = -4.75 - 2.27*(X[1] - 0.3) + 6.0;
  }

  return value;
}
```

Figure 6.7c *C Function* for *the linear inequality constraints of A Model*

6.3 A User Interface and Interactive Decision Making

6.3.1 Menu-Driven Interfaces

The user of the IMC-DSS may use a rule-based advisory system to select an appropriate MCDM method. In this system, a number of questions with several possible answers have been designed and the user can select a particular MCDM method by responding to these questions. Alternatively, he may directly choose a MCDM method from menus listing all the MADM and MODM methods available, as shown in Figure 6.9.

If a MCDM method is chosen, then the menu-driven interface for the selected method will be initialized to support preference acquisition and interactive decision making. For instance, Figures 6.10 and 6.11 demonstrate the interface designed for the ISTM method, which is composed of three parts [Yang 1992b]. Part 1 is a table which is designed to support decision analysis by listing the best, worst and current achievement levels of objectives. The first column in the table explains whether an objective is for maximization or for minimization. In the third and fifth columns, the best and worst values of objectives are listed. The "best" value here means that it is the optimal feasible value of an objective

```
/*
** The list of nonlinear equality constraint functions
*/

double nl_eqs_def(double *X, int i, int id_model)
{

  double value;

  /* Multiple objective model 15 */
  if(id_model == 15)
  {
    if(i == 1)
      value = 0;
  }

  return value;
}
```

Figure 6.7d *C Function* for *the nonlinear equality constraints of A Model*

while the "worst" value is the smallest value of a benefit objective or the largest value of a cost objective in a pay-off table (see Table 4.2) obtained while searching for the best values of the objectives. The fourth column displays the values of objectives at the newly generated efficient (feasible) design. The last column illustrates the relative achievement levels of the new design on objectives compared to the ideal design (generally infeasible).

The second and the third parts of the interface present the trade-off questions designed in the ISTM method, as headed by "CONDUCT TRADEOFF ANALYSIS" and "INDICATE THE ABSOLUTE DECREMENTS", respectively. The second part is used to support the classification of the set of objectives into three subsets, as discussed in Section 4.3.4. The user can provide a tradeoff direction by responding to the questions. The third part is used to assist in the assignment of step sizes for those objectives earmarked for sacrifice.

The interfaces for preference acquisition and interactive decision making for the other MCDM methods have also been designed with their own features. Through these interfaces, the user is able to use the IMC-DSS to deal with his decision problem even though he may not know the details of the mathematics involved

```
/*
** The list of nonlinear inequality constraint functions
*/

double nl_ineqs_def(double *X, int i, int id_model)
{
  double value;

  /* Multiple objective model 15 */
  if(id_model == 15)
  {
    if(i == 1)
      value = 6.0 - 5.1 - 0.177*(X[1] - 0.3) - 0.978*(X[2] - 0.3)
            - 0.216*(0.39/(1.39 - X[1]*X[1]) - 0.3)
            - 0.768*(0.39/(1.39 - X[2]*X[2]) - 0.3);
    if(i == 2)
      value = 2.50*pow(10.0, -3.0)*(450.0/(1.09 - X[3]*X[3]) - 450.0) - 1.5;
    if(i == 3)
      value = 3.5 - 1.0 - 0.0332*(X[1] - 0.3) - 0.0186*(X[2] - 0.3)
            - 3.34*(X[3] - 0.3) - 0.0204*(0.39/(1.39 - X[1]*X[1]) - 0.3)
            - 0.78*(0.39/(1.39 - X[2]*X[2]) - 0.3)
            - 2.62*(0.39/(1.39 - X[3]*X[3]) - 0.3);
  }

  return value;
}
```

Figure 6.7e *C Function* for *the nonlinear inequality constraints of A Model* in the MCDM methods.

6.3.2 A Unified Approach for Generating and Ranking Designs

In the IMC-DSS, the user may choose one particular MCDM method to solve his problem. He may also use multiple methods to investigate the same problem from different angles. Such an investigation can help the user to better understand the problem in hand. Multiple MCDM methods may be used for problem solving in the following two ways.

Input_data_file_for_recording_the_structural_information_of_model_15

Identifier_of_the_model	15
Number_of_objective_functions	3
Number_of_design_variables	3
Number_of_linear_equality_constraints	0
Number_of_linear_inequality_constraints	1
Number_of_nonlinear_equality_constraints	0
Number_of_nonlinear_inequality_constraints	3
Identifier_for_printing_iteration_information	1
Identifier_of_the_method_for_calculating_derivatives	1
Maximum_iteration_number_for_nonlinear_optimization	200
Initial_design_variable_value_X[1]	1.0
Initial_design_variable_value_X[2]	1.0
Initial_design_variable_value_X[3]	0.5
Maximum_design_variable_value_X[1]	1.0
Maximum_design_variable_value_X[2]	1.0
Maximum_design_variable_value_X[3]	1.0
Minimum_design_variable_value_X[1]	0.3
Minimum_design_variable_value_X[2]	0.3
Minimum_design_variable_value_X[3]	0.3
Initial_step_size_for_variable_X[1]	0.2
Initial_step_size_for_variable_X[2]	0.2
Initial_step_size_for_variable_X[3]	0.2

Figure 6.8 *The Data File* for *A Model*

Firstly, a MADM problem (or MODM problem) may be investigated using different MADM methods (or MODM methods). This strategy is useful as the MADM (or MODM) methods of the IMC-DSS can provide different ways of acquiring and representing user preference information which governs the selection of the best compromise designs. Through the use of multiple methods, various features of a problem can be examined. Consequently, the user could

**************************************	**************************************
MADM Methods	MODM Methods
(Multiple Attribute Decision Making)	(Multiple Objective Decision Making)
Engineering Design Centre	Engineering Design Centre
University Of Newcastle Upon Tyne	University Of Newcastle Upon Tyne
**************************************	**************************************
A = generation of weights for attributes	A = single objective optimization
B = TOPSIS method	B = minimax (ideal point) method
C = CODASID method	C = interactive step trade-off method(ISTM)
D = AHP method	D = goal programming
E = UTA method	E = Geoffrion's method
Q = Quit	Q = Quit
Please choose a method(A/B/C/D/E/Q)__	Please choose a method(A/B/C/D/E/Q)__

Figure 6.9 *The Menu* for *direct selection of MADM* and *MODM Methods*

gain a better insight into the problem and may thus be in a better position to choose his best compromise design.

Secondly, a MODM problem may be dealt with using both MODM and MADM methods. For instance, we can use an interactive MODM method, such as the ISTM method or Geoffrion's method, to generate a set of "good efficient designs" from which a good compromise design may be evolved using a MADM method such as the TOPSIS or CODASID method. This strategy is useful as it is generally rather difficult to construct for a MODM problem an overall preference structure such as a utility function for ranking alternative designs without subscribing to some restrictive assumptions such as the preferential independence assumption about criteria for an additive utility function. In engineering design, however, most such assumptions may not always be acceptable. If a sufficiently large number of typical designs are generated, this strategy can lead to the evolution of a good final design using a range of multiple MADM methods.

The IMC-DSS aims to integrate several MCDM methods with different features into a single system and hence provides an appropriate environment for implementing the above strategies. A few examples will be examined in the following sections to demonstrate the application of the IMC-DSS in design selection and synthesis.

THE DECISION ANALYSIS TABLE

		(a)best value	(c)current value	(b)worst value	rate $\frac{(c-b)}{(a-b)}$
MAX	f1	36.06	31.48	21.87	67.74 %
MAX	f2	1.30410e+04	8448.19	2288.60	57.29 %
MAX	f3	3.02486e+04	2.01866e+04	7608.83	55.56 %
MAX	f4	32.29	26.11	18.33	55.75 %
MAX	f5	32.32	26.09	18.31	55.56 %
MIN	f6	4.80573e+06	9.45044e+06	1.52565e+07	55.56 %

CONDUCT TRADE-OFF ANALYSIS

Please classify the objectives into one of the following types

1. An objective has to be IMPROVED from the current status (I).
2. An objective must be KEPT at or above the current status (K).
3. An objective may be SACRIFICED from the current status (S).

Objective 1 (I or K or S)__k
Objective 2 (I or K or S)__s
Objective 3 (I or K or S)__s
Objective 4 (I or K or S)__s
Objective 5 (I or K or S)__s
Objective 6 (I or K or S)__i

Are you sure that you classified the objectives properly (Y or N)?___y

Figure 6.10 *The First Part of the Interface* for *ISTM*

6.4 Application of IMC-DSS

6.4.1 A Multiattribute Vessel Choice Problem

A vessel choice problem of selecting one type of ships from three candidate vessel types, that is, general cargo type, ro-ro type and full-container type, was shown in Figure 3.14 and examined in Section 3.3.1 [Sen 1992]. In this section, the same problem is re-examined to demonstrate how to use the interface of the IMC-DSS for entering raw data and obtain the rankings of alternative designs.

In the IMC-DSS, the evaluation framework and all the pairwise comparison matrices of the problem are stored in a data file, as shown in Figure 6.6, which can be created using the menu-driven interface designed for the AHP method. For instance, Tables 3.6-1 and 3.6-2 can be entered into the data file in a way as shown in Figure 6.12 [Yang 1992a].

THE DECISION ANALYSIS TABLE

		(a)best value	(b)current value	(c)worst value	rate $\frac{(c-b)}{(a-b)}$
MAX	f1	36.06	31.48	21.87	67.74 %
MAX	f2	1.30410e+04	8448.19	2288.60	57.29 %
MAX	f3	3.02486e+04	2.01866e+04	7608.83	55.56 %
MAX	f4	32.29	26.11	18.33	55.75 %
MAX	f5	32.32	26.09	18.31	55.56 %
MIN	f6	4.80573e+06	9.45044e+06	1.52565e+07	55.56 %

INDICATE THE ABSOLUTE DECREMENTS

Please indicate the absolute value by which the objective can be sacrificed

the decrement of objective 2 = 1000

the decrement of objective 3 = 2000

the decrement of objective 4 = 1

the decrement of objective 5 = 1

Are you sure that these assigned values are adequate (Y or N)?___y

Figure 6.11 *The Second Part of Interface* for *ISTM*

In the IMC-DSS, not only Saaty's standard AHP method but two extensions are used as well to produce three sets of relative weights and rankings for all elements (attributes, alternatives or decision makers) at a single level. The first extension of AHP was proposed by Belton and Gal. using a different scheme to normalize eigenvectors [Belton and Gal 1981]. Johnson et al. presented the second extension using left eigenvectors, instead of right eigenvectors, as weights [Johnson, Beine and Wang 1979]. These three sets of weights and rankings may be referred to as Saaty's, Belton's and Johnson's weights and rankings respectively. Belton's and Johnson's rankings are used to check the consistency of the pairwise comparisons. If the three rankings are the same, the consistency of the pairwise comparisons is accepted. In this case, Saaty's weights may be used to represent the relative importance of the corresponding elements. If consistency is unsatisfactory, the IMC-DSS will suggest either

(1) to revise those pairwise comparison matrices, the consistency indices of which are unsatisfactory, or

(2) to use Saaty's weight and ranking as the final result.

Table 6.1 shows the results obtained for the problem. Obviously, Saaty's and Belton's rankings for the vessel types are the same but different from Johson's ranking.

****

The following rules can help you to quantify your pairwise comparisons

A(i)	Equally Important	A(j),	then A(i)/A(j)= 1 or A(j)/A(i)=-1
A(i)	Weakly More Important	A(j),	then A(i)/A(j)= 3 or A(j)/A(i)=-3
A(i)	Moderately More Important	A(j),	then A(i)/A(j)= 5 or A(j)/A(i)=-5
A(i)	Strongly More Important	A(j),	then A(i)/A(j)= 7 or A(j)/A(i)=-7
A(i)	Very Strongly More Important	A(j),	then A(i)/A(j)= 9 or A(j)/A(i)=-9

****

Please enter PAIRWISE comparisons for ATTRIBUTES at level II _ 3

Attribute(1,3) / Attribute(2,3) = 1
Attribute(1,3) / Attribute(3,3) = 1
Attribute(2,3) / Attribute(3,3) = 1

Are you sure you entered correct data ?(y/n)
(If Y, go on; if N, enter the data again)---y

Please enter PAIRWISE comparisons for the ATTRIBUTES at level II _ 2
on the basis of ATTRIBUTE 1 at level II _ 3

Attribute(1,1,3) / Attribute(2,1,3) = -2
Attribute(1,1,3) / Attribute(3,1,3) = 1
Attribute(1,1,3) / Attribute(4,1,3) = -3
Attribute(1,1,3) / Attribute(5,1,3) = 2
Attribute(2,1,3) / Attribute(3,1,3) = -2
Attribute(2,1,3) / Attribute(4,1,3) = -3
Attribute(2,1,3) / Attribute(5,1,3) = 2
Attribute(3,1,3) / Attribute(4,1,3) = 1
Attribute(3,1,3) / Attribute(5,1,3) = 3
Attribute(4,1,3) / Attribute(5,1,3) = 2

Are you sure you entered correct data ?(y/n)
(If Y, go on; if N, enter the data again)---y

Figure 6.12 *The Interface* for *Entering Data* for *AHP*

If Saaty's weight and ranking are used as the final results, full-container vessels will be the best choice. Since the consistency is not absolute, however, it may be desirable to revise those pairwise comparison matrices whose consistency is unsatisfactory. The IMC-DSS can provide a consistency index for checking the consistency of a pairwise comparison matrix. The consistency indices of all the comparison matrices of the problems as defined by Tables 3.6-1 3.6-27 are as

Table 6.1 *Results for The Vessel Choice Problem*

	Saaty's		Belton's		Johson's	
elements	weight	ranking	weight	ranking	weight	ranking
d_1	0.3333	1	0.3333	1	0.3333	1
d_2	0.3333	1	0.3333	1	0.3333	1
d_3	0.3333	1	0.3333	1	0.3333	1
f_1	0.2471	1	0.6764	1	0.1482	1
f_2	0.2396	2	0.6475	2	0.1546	2
f_3	0.1771	3	0.4992	3	0.2051	3
f_4	0.1669	5	0.4853	4	0.2614	5
f_5	0.1692	4	0.4650	5	0.2307	4
y_1	0.0989	1	0.2439	1	0.0581	11
y_2	0.0989	1	0.2439	1	0.0581	11
y_3	0.0494	8	0.1219	8	0.0605	15
y_4	0.0293	11	0.0722	10	0.0586	12
y_5	0.0949	2	0.2335	2	0.0560	8
y_6	0.0274	12	0.0674	12	0.0589	13
y_7	0.0880	3	0.2167	3	0.0560	9
y_8	0.0250	13	0.0654	13	0.0562	10
y_9	0.0584	7	0.1530	7	0.0536	6
y_{10}	0.0250	13	0.0654	13	0.0562	10
y_{11}	0.0687	6	0.1800	4	0.0533	4
y_{12}	0.0323	10	0.0714	11	0.0518	3
y_{13}	0.0790	4	0.1750	5	0.0493	1
y_{14}	0.0161	15	0.0357	15	0.0559	7
y_{15}	0.0395	9	0.0875	9	0.0510	2
y_{16}	0.0725	5	0.1677	6	0.0536	11
y_{17}	0.0725	5	0.1677	6	0.0536	11
y_{18}	0.0242	14	0.0559	14	0.0593	2
a_1	0.2603	3	0.4756	3	0.3683	3
a_2	0.3598	2	0.6652	2	0.3013	1
a_3	0.3799	1	0.6851	1	0.3303	2

shown in Table 6.2. A threshold value for consistency index may be used to decide if the consistency of a comparison matrix is acceptable. If 0.05 is used as such a threshold, for example, the pairwise comparison matrices 2, 3, 7, 18 and 22 are required to be modified to improve the consistency.

Table 6.2 *Consistency Indices of Comparison Matrices*

matrix	value	matrix	value	matrix	value
1	0.000000	10	0.004601	19	0.004601
2	0.072451	11	0.004601	20	0.004601
3	0.084894	12	0.004601	21	0.004601
4	0.008279	13	0.004601	22	0.067806
5	0.000000	14	0.026810	23	0.000000
6	0.003454	15	0.004601	24	0.004601
7	0.051500	16	0.026810	25	0.000000
8	0.013747	17	0.004601	26	0.026811
9	0.000000	18	0.067805	27	0.004601

6.4.2 A Multiobjective Semi-Submersible Design Problem

A preliminary design model for semi-submersibles (SS-model) has been built as a multiple objective decision making problem, as described in Section 3.2.4.3. For this new design problem, it is very difficult to establish an overall preference structure, either a utility function or a set of target values, for the six objective functions without comprehensive investigation of the feasible design space. The ISTM method may be employed for investigating the efficient design frontier of the SS-model.

ISTM first optimizes each of these objectives to generate the extreme designs for this problem. Table 6.3 shows the values of the objectives at the generated extreme designs, where a_i is obtained by optimizing the ith objective. The values of the design variables at the extreme designs are as shown in Table 3.4.

Then ISTM searches for an efficient design called the ideal point design using the minimax method. The values of the objectives obtained at this design are shown in the fourth column of the decision analysis table in Figure 6.10. The design obtained using the minimax method may be regarded as a good design as it is the closest feasible design, in some sense, to the ideal design taking the best values of all the objectives. If the designer is not satisfied with this design, however, he can carry out further search for better designs by means of trade-off analysis among the objectives.

It is quite convenient to conduct trade-off analysis by using the interface designed for the ISTM method in the IMC-DSS, as shown in Figures 6.13 to 6.17 [Yang 1992b].

Table 6.3 *The Pay off Table for The Semi–Submersible Model*

extreme designs	objective values					
	y_1	y_2	y_3	y_4	y_5	y_6
a_1	36.06	9594.95	18014.24	18.33	18.31	7594448.00
a_2	32.64	13040.96	26194.62	21.33	21.34	12404110.00
a_3	31.26	12284.91	30248.64	25.92	25.93	15256452.00
a_4	28.21	6952.27	20811.53	32.29	32.32	13551909.00
a_5	28.23	6748.00	20270.23	32.29	32.32	13205347.00
a_6	21.87	2288.60	7608.83	20.36	20.35	4805733.50

In Figure 6.13, one can find that the cost of steel construction at the current design is over 8.77 million pounds. Suppose the designer's target is to keep the cost below 8 millions with all the other objectives being achieved as much as possible. The tradeoff analysis at the current design (a_7), as shown in Figure 6.13, reads that objective 6 needs to be improved (reduced) and objective 3 kept at its current level at the expense of objectives 1, 2, 4 and 5. The decrement of the four objectives earmarked for sacrifice are 1.48, 1278.76, 0.11 and 0.1, respectively. An auxiliary problem embedding the above analysis is then constructed and solved, resulting in a new design denoted by a_8. The objective values at a_8 are shown by the fourth column of the decision analysis table in Figure 6.14. The cost has been reduced to just over 8.5 millions, still above the target level.

Similar tradeoff analysis can be conducted with cautious assignment of decrements for objectives designated for sacrifice, as shown in Figures 6.14, 6.15, 6.16 and 6.17, until the cost is reduced below the target value of 8 millions. This sequence of tradeoff analyses are significant in the sense that the target cost should be achieved by properly sacrificing the other objectives in a sensible way. In other words, unnecessary sacrifice of any objectives should always be avoided. The values of the objectives and the design variables at the designs generated in the tradeoff analyses are listed in Tables 6.4 and 6.5.

THE DECISION ANALYSIS TABLE

		(a)best value	(c)current value	(b)worst value	rate $\frac{(c-b)}{(a-b)}$
MAX	f1	36.06	31.48	21.87	67.74 %
MAX	f2	1.30410e+04	8278.76	2288.60	55.71 %
MAX	f3	3.02486e+04	1.81866e+04	7608.83	46.72 %
MAX	f4	32.29	25.11	18.33	48.58 %
MAX	f5	32.32	25.10	18.31	48.46 %
MIN	f6	4.80573e+06	8.77129e+06	1.52565e+07	62.05 %

CONDUCT TRADE-OFF ANALYSIS

Please classify the objectives into one of the following types

1. An objective has to be IMPROVED from the current status (I).
2. An objective must be KEPT at or above the current status (K).
3. An objective may be SACRIFICED from the current status (S).

Objective 1 (I or K or S)__s
Objective 2 (I or K or S)__s
Objective 3 (I or K or S)__k
Objective 4 (I or K or S)__s
Objective 5 (I or K or S)__s
Objective 6 (I or K or S)__i

Are you sure that you classified the objectives properly (Y or N)?___y

INDICATE THE ABSOLUTE DECREMENTS

Please indicate the absolute value by which the objective can be sacrificed

the decrement of objective 1 = 1.48
the decrement of objective 2 = 1278.76
the decrement of objective 4 = 0.11
the decrement of objective 5 = 0.1

Are you sure that these assigned values are adequate (Y or N)?___y

Figure 6.13 *The Interface of ISTM Showing Tradeoff Analysis at* a_7

Once the C functions and the data file defining the model are provided in the way discussed in Section 6.2.1, the above tradeoff analysis can be conducted by simply responding to the questions shown in the interface designed for the ISTM method. All the relevant calculations involved in ISTM can then be carried out

THE DECISION ANALYSIS TABLE

		(a)best value	(c)current value	(b)worst value	rate $\frac{(c-b)}{(a-b)}$
MAX	f1	36.06	30.45	21.87	60.52 %
MAX	f2	1.30410e+04	7475.56	2288.60	48.24 %
MAX	f3	3.02486e+04	1.81866e+04	7608.83	46.72 %
MAX	f4	32.29	25.02	18.33	47.94 %
MAX	f5	32.32	25.00	18.31	47.75 %
MIN	f6	4.80573e+06	8.50312e+06	1.52565e+07	64.62 %

CONDUCT TRADE-OFF ANALYSIS

Please classify the objectives into one of the following types

1. An objective has to be IMPROVED from the current status (I).
2. An objective must be KEPT at or above the current status (K).
3. An objective may be SACRIFICED from the current status (S).

Objective 1 (I or K or S)___s
Objective 2 (I or K or S)___s
Objective 3 (I or K or S)___s
Objective 4 (I or K or S)___s
Objective 5 (I or K or S)___k
Objective 6 (I or K or S)___i

Are you sure that you classified the objectives properly (Y or N)?___y

INDICATE THE ABSOLUTE DECREMENTS

Please indicate the absolute value by which the objective can be sacrificed

the decrement of objective 1 = 1.45
the decrement of objective 2 = 475.56
the decrement of objective 3 = 2186.6
the decrement of objective 4 = 0.02

Are you sure that these assigned values are adequate (Y or N)?___y

Figure 6.14 *The Interface of ISTM Showing Tradeoff Analysis at* a_8

automatically in the sense that the user does not need to construct or solve the corresponding auxiliary problems, which may not be a trivial task. The above menu-driven interface designed for ISTM provides a general means for facilitating tradeoff analysis although other types of interfaces such as graphic

THE DECISION ANALYSIS TABLE

		(a)best value	(c)current value	(b)worst value	rate $\frac{(c-b)}{(a-b)}$
MAX	f1	36.06	29.61	21.87	54.53 %
MAX	f2	1.30410e+04	6999.99	2288.60	43.82 %
MAX	f3	3.02486e+04	1.60000e+04	7608.83	37.06 %
MAX	f4	32.29	25.01	18.33	47.85 %
MAX	f5	32.32	25.00	18.31	47.75 %
MIN	f6	4.80573e+06	8.14608e+06	1.52565e+07	68.04 %

CONDUCT TRADE-OFF ANALYSIS

Please classify the objectives into one of the following types

1. An objective has to be IMPROVED from the current status (I).
2. An objective must be KEPT at or above the current status (K).
3. An objective may be SACRIFICED from the current status (S).

Objective 1 (I or K or S)___s
Objective 2 (I or K or S)___s
Objective 3 (I or K or S)___i
Objective 4 (I or K or S)___s
Objective 5 (I or K or S)___s
Objective 6 (I or K or S)___i

Are you sure that you classified the objectives properly (Y or N)?___y

INDICATE THE ABSOLUTE DECREMENTS

Please indicate the absolute value by which the objective can be sacrificed

the decrement of objective 1 = 0.61
the decrement of objective 2 = 500
the decrement of objective 4 = 1.01
the decrement of objective 5 = 1

Are you sure that these assigned values are adequate (Y or N)?___y

Figure 6.15 *The Interface of ISTM Showing Tradeoff Analysis at* a_9

interfaces could also be designed.

Similar interfaces have been designed for the other MCDM methods in the IMC-DSS and may be employed to investigate the same problem, as discussed in

THE DECISION ANALYSIS TABLE

		(a)best value	(c)current value	(b)worst value	rate $\frac{(c-b)}{(a-b)}$
MAX	f1	36.06	30.70	21.87	62.23 %
MAX	f2	1.30410e+04	7438.29	2288.60	47.89 %
MAX	f3	3.02486e+04	1.79011e+04	7608.83	45.46 %
MAX	f4	32.29	24.02	18.33	40.79 %
MAX	f5	32.32	24.00	18.31	40.61 %
MIN	f6	4.80573e+06	8.14608e+06	1.52565e+07	68.04 %

CONDUCT TRADE-OFF ANALYSIS

Please classify the objectives into one of the following types

1. An objective has to be IMPROVED from the current status (I).
2. An objective must be KEPT at or above the current status (K).
3. An objective may be SACRIFICED from the current status (S).

Objective 1 (I or K or S)___s
Objective 2 (I or K or S)___s
Objective 3 (I or K or S)___s
Objective 4 (I or K or S)___k
Objective 5 (I or K or S)___k
Objective 6 (I or K or S)___i

Are you sure that you classified the objectives properly (Y or N)?___y

INDICATE THE ABSOLUTE DECREMENTS

Please indicate the absolute value by which the objective can be sacrificed

the decrement of objective 1 = 1.7
the decrement of objective 2 = 238.29
the decrement of objective 3 = 901

Are you sure that these assigned values are adequate (Y or N)?___y

Figure 6.16 *The Interface of ISTM Showing Tradeoff Analysis at* a_{10}

the following section.

THE DECISION ANALYSIS TABLE

		(a)best value	(c)current value	(b)worst value	rate $\frac{(c-b)}{(a-b)}$
MAX	f1	36.06	30.28	21.87	59.29 %
MAX	f2	1.30410e+04	7199.99	2288.60	45.68 %
MAX	f3	3.02486e+04	1.70001e+04	7608.83	41.48 %
MAX	f4	32.29	24.02	18.33	40.79 %
MAX	f5	32.32	24.00	18.31	40.64 %
MIN	f6	4.80573e+06	7.99342e+06	1.52565e+07	69.50 %

CONDUCT TRADE-OFF ANALYSIS

Please classify the objectives into one of the following types

1. An objective has to be IMPROVED from the current status (I).
2. An objective must be KEPT at or above the current status (K).
3. An objective may be SACRIFICED from the current status (S).

Objective 1 (I or K or S)__k
Objective 2 (I or K or S)__k
Objective 3 (I or K or S)__k
Objective 4 (I or K or S)__k
Objective 5 (I or K or S)__k
Objective 6 (I or K or S)__k

Are you sure that you classified the objectives properly (Y or N)?__y

Figure 6.17 *The Interface of ISTM Showing Tradeoff Analysis at* a_{11}

Table 6.4 *Objective Values of Efficient Designs* for *The SS_Model*

efficient designs	objective values					
	y_1	y_2	y_3	y_4	y_5	y_6
a_7	31.479	8278.758	18186.627	25.108	25.099	8771294.00
a_8	30.455	7475.557	18186.623	25.019	24.999	8503116.00
a_9	29.605	6999.994	16000.020	25.006	24.999	8146078.00
a_{10}	30.698	7438.285	17901.070	24.020	23.999	8146078.00
a_{11}	30.280	7199.994	17000.068	24.020	24.003	7993421.00

Table 6.5 *Variable Values of the Generated Efficient Designs*

design variables	efficient designs				
	a_7	a_8	a_9	a_{10}	a_{11}
x_1	13.76	13.66	13.27	13.43	13.27
x_2	7.00	7.00	7.00	7.00	7.00
x_3	39.67	38.15	38.72	36.08	36.25
x_4	22.73	21.40	20.72	21.46	21.10
x_5	9.356	10.00	9.28	10.00	9.74
x_6	117.28	116.57	113.62	114.82	113.61
x_7	4.00	4.00	4.00	4.00	4.00
x_8	82.55	81.99	79.65	80.60	79.64
x_9	1.00	1.00	1.00	1.00	1.00

6.4.3 Design Using the Unified Approach

Let us suppose all the designs (a_7 to a_{11}) generated in the above tradeoff analyses are acceptable to the designer but he is not certain which one might be his best choice although he does have the ability of assigning weights to the objectives. Then, he may either use one of the MADM methods to select the best design from a_7 to a_{11} or employ other MODM methods to search for other compromise designs based on the efficient designs already generated.

Suppose the designer chooses the CODASID method to rank a_7 to a_{11}. The decision table can be constructed as shown in Table 6.4. Two value systems are tested as shown in Table 6.6. In value system 1, it is assumed that f_1, f_2, f_3, f_4 and f_5 are of equal importance and f_6 is twice as important as the other five objectives. Secondly, suppose f_6 is overwhelmingly (9 times) more important than f_1 to f_5 which are all equally important.

Table 6.6 *Two Value Systems* for *Selection of The Efficient Designs*

value systems	weights for objectives					
	y_1	y_2	y_3	y_4	y_5	y_6
1	1	1	1	1	1	2
2	1	1	1	1	1	9

Following the computational steps of the CODASID method, these five efficient designs can be ranked as shown in Table 6.7. For value system 1, a_9, instead of a_{11}, is selected as the best compromise design as at a_9 cost is quite low and other objectives also reach good achievement levels.

Table 6.7 *The Rankings Generated Using The CODASID Method*

value systems	rankings of efficient designs				
	a_7	a_8	a_9	a_{10}	a_{11}
1	5	3	1	2	4
2	5	4	2	3	1

Suppose the designer is not satisfied with any of the generated designs. If, as a result of the above tradeoff analyses, he is now able to assign target values for all the objectives, then the goal programming method may be used to search for the best compromise design.

Figure 6.18 illustrates how to use the interface designed for the goal programming method in the IMC-DSS to search for such a compromise design. There are four parts in the interface. In the first part, the target (goal) values for the objectives can be entered at the keyboard as shown in the fifth column. If the priority of any objectives is absolutely higher than that of others, this can be declared in the second part. In the last part of the interface, the user can indicate how his target for each of the objectives should be achieved, that is whether an objective should be under-achieved, over-achieved or achieved exactly. The above preference information is required by the goal programming method.

Once the above preference information is provided, a goal programming model is automatically formulated and solved by the IMC-DSS. Table 6.8 lists the values of the design variables and the objectives at the newly generated efficient design a_{12}. It is obvious that both a_{11} and a_{12} reach the same level of construction cost but they are different from each other in the achievement levels of the other objectives.

PLEASE ASSIGN THE GOAL VALUE TO EVERY OBJECTIVE

		The best value	The worst value	The goal value
MAX	f1	36.06	21.87	31.48
MAX	f2	1.30410e+04	2288.60	8278.76
MAX	f3	3.02486e+04	7608.83	18186.63
MAX	f4	32.29	18.33	25.11
MAX	f5	32.32	18.31	25.10
MIN	f6	4.80573e+06	1.52565e+07	7993421.00

Are you sure that these assigned goal values are adequate(Y or N)?__y

Is there any objective absolutely more important than another?
Or, are the objectives on different priority levels(Y or N)?___n

Do you think that the 6 objectives are of EQUAL importance(Y or N)?___n

Please set a VALUE to an objective for the relative weight
The weight VALUE of objective 1 = 1
The weight VALUE of objective 2 = 1
The weight VALUE of objective 3 = 1
The weight VALUE of objective 4 = 1
The weight VALUE of objective 5 = 1
The weight VALUE of objective 6 = 9

Are you sure that the assigned values are adequate (Y or N)?___y

Please indicate whether an objective should be attained

<1> as CLOSELY TO the goal value as possible; or
<2> as ABOVE the goal value as possible; or
<3> as UNDER the goal value as possible.

The attainment level of objective 1 <1 or 2 or 3> = 1
The attainment level of objective 2 <1 or 2 or 3> = 1
The attainment level of objective 3 <1 or 2 or 3> = 1
The attainment level of objective 4 <1 or 2 or 3> = 1
The attainment level of objective 5 <1 or 2 or 3> = 1
The attainment level of objective 6 <1 or 2 or 3> = 1

Are you sure that these attainment levels are adequate(Y or N)?___y

Figure 6.18 *The Interface of Goal Programming* for *Generating* a_{12}

Table 6.8 *Results Generated Using Goal Programming*

objectives		design variables	
symbol	value	symbol	value
f_1	31.48	x_1	13.35
f_2	7767.42	x_2	7.00
f_3	18186.63	x_3	34.54
f_4	23.15	x_4	22.12
f_5	23.13	x_5	9.96
f_6	7993421.50	x_6	114.15
		x_7	4.00
		x_8	80.07
		x_9	1.00

6.5 Concluding Comments

The methods examined in Chapters 3 and 4 are embedded in a decision support system that is loosely structured, allowing for a large range of user interactions and provides the opportunity for future growth. It is considered that the use of MCDM for practical decision making in engineering design or in any other domain requires the assistance of some such system, so that the user is provided with the necessary backup in the use of the analytical tools.

The decision support environment continues to benefit from the work within the Decision Support Group at Newcastle under Prof. Sen. New methods and features are being added on and the experience gained in the use of these methods continues to inform the research into this area of work for both the authors.

7
Past, Present and the Future

7.1 Introduction

The preceding chapters set out the benefits associated with the application of MCDM in engineering design and examine how the use of different methods leads to results of a correspondingly distinctive nature. The data requirements for the various approaches are also discussed along with the underlying assumptions that are built in. As in all decision making methodologies there are two principal considerations in doing this. These are

(i) how can one make rational decisions in a consistent manner when faced with multiple, conflicting criteria?

(ii) what do real decision-makers do when confronted with decisions involving multiple, conflicting criteria?

The first of the two considerations addresses the normative issues of decision making. The aim is to establish more or less standard approaches to broad classes of problems so that the decisions are consistently arrived at. This has the added virtue of allowing computer based decision support tools to be prepared with some degree of confidence in that they will be used in a generally predictable manner. The second of the considerations above deals with the descriptive aspects of decision-making. In other words it is largely about the capturing of the rules applied by real DMs when confronted with real problems.

It is obviously desirable that the corresponding normative and descriptive approaches be as close to each other as possible as this adds to the credibility of them both. A good normative method should not be at variance with the observed behaviour of real DMs if it is to be relevant. Good practical decision making, on the other hand, involves information processing of the kind that benefits from a systematic normative approach. In view of the above the position of a rational DM may well be to use appropriate MCDM methodologies only to clarify the decision situation with a view to taking more informed decisions using more familiar methods.

Another important consideration is the distance that is brought about between a statement of a decision problem and its solution by the processing that is involved in the use of any methodology. Intelligent assessment of a problem necessarily involves processing of the basic data in one form or another. But unless this is carefully done the processing may result in a loss of "feel" for the problem on the part of the DMs. Most practical decision makers are aware of this and often choose methods that are capable of addressing the problem at hand without obscuring some of the more subtle aspects of the choices available. In order to address this consideration it is sometimes also desirable to try out more than one method, so that any bias introduced by one method may be counter-balanced by the insights provided by another.

It follows from the above, therefore, that a decision making method should only be used if the underlying assumptions of the method are understood. This is why it is often better to use a simple method where the underlying assumptions are easier to spot and cater for than a more complicated procedure. The choice of a multiple criteria decision making method is thus itself a multiple criteria problem.

Finally, most real life problems are not totally capable of being processed by one method and one method only. Such problems are often stated in a form that requires a hybrid processing strategy, where a range of methodologies, not all of which are specifically MCDM methods, need to be used. It would be helpful to end this text with a couple of examples of such hybrid solution strategies as examples.

7.2 Case Studies

7.2.1 Designing product development processes to minimise lead times.

In recent years, as a result of rapid technological change, increased market segmentation and reduced product life cycles, time to market has increasingly become a leading focus for gaining competitive advantage. Time to market, on the other hand, has been dominated by product development lead times. It is thus extremely important today to plan and schedule product development activities with the specific aim of reducing product development lead times.

In pursuing the above it is obviously crucial to be able to represent and manipulate sequences of product development activities. One systematic way of doing so is to use the design structure matrix (DSM) system [Steward 1981]. This is a flexible tool and has been developed in recent years by Gebala and Eppinger [Gebala & Eppinger 1991], Kusiak and Park [Kusiak & Park 1990], Rogers [Rogers 1997] and Austin et al [Austin 1996]. Most applications so far, however, reported in literature have dealt with the development of procedures and resequencing strategies leading to minimisation of iteration in the belief that minimum iteration results inevitably in lower lead times. However, it can be shown that minimisation of iterations in product development activities often leads to simultaneous reductions in opportunities for parallel or concurrent activities. Designing product development processes using a multiple criteria decision making approach that combines the twin objectives of minimising iteration and maximising concurrency, and indeed other objectives, if appropriate, is thus a more robust and satisfactory approach compared to the use of the single objective of minimising iteration.

Design Structure Matrix (DSM) Systems

Efficient planning and scheduling of product development work inevitably involves data inter-relationships of a kind that often forces iterations. This is because a given activity is often influenced by the results of some downstream activities, and these results must be taken into account when they become available in reassessing the given activity. Conventional planning tools like network analysis techniques do not, however, cater for iterations, so an alternative representation and manipulation tool is necessary. The DSM system offer the flexibility that is sought here. It does this through a matrix representation of data dependencies as shown in Figure 7.1

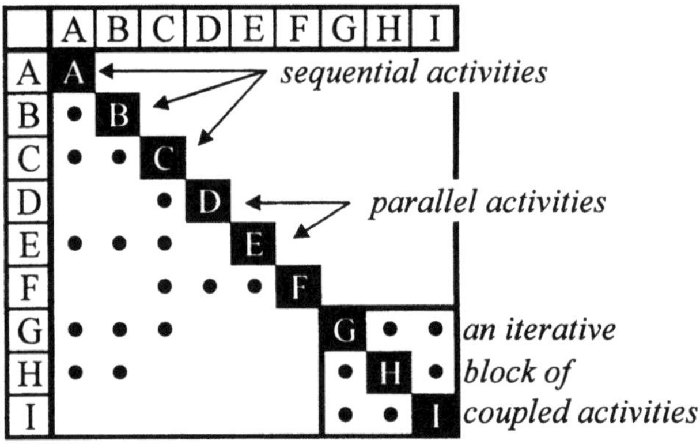

Figure 7.1 *A Design Structure Matrix*

In the above matrix, activities are listed in rows and columns in the sequence in which they are executed. Reading across a row it is possible to identify all of those activities whose output data is required to perform the activity corresponding to the row in question, whereas reading down a column indicates all of the activities that receive the output data from the activity corresponding to the column. Thus in Figure 7.1, activity G receives the output from H and I whereas activity A passes its output to activities B, C, E, G and H.

The relative positioning of the data dependency marks (represented by dots in this figure) identifies the type of relationship that links activities; *sequential, parallel* or *coupled*. Although in this example only a binary representation is modelled (i.e. only Yes/No in terms of data dependency) there is the opportunity to use scaled values, if they can be estimated, to quantify the relative impact of the different data dependencies. In the example shown in Figure 7.1 activity A passes data to activity B, amongst other downstream activities, and B passes data to activity C, amongst other activities. Hence activities A, B and C are sequential. In a similar manner it is easy to see that, as activity D neither receives data from nor sends data to activity E, these two activities may be pursued in parallel, if other conditions, like availability of resources, allow this to happen. On the basis of the above arguments it is clear that activities G, H and I both receive data from and send data to each other and are thus coupled.

Minimising Iteration

Based on the matrix representation described above, new scheduling strategies can be derived by re-sequencing activities based on some predefined objective or objectives. This can be done using single or multiple objective genetic algorithms, for example. Figure 7.2(a) shows an input matrix of activities and data dependencies where the strength of the relationships are represented by the descriptors *Strong* (S), *Medium* (M) and *Weak* (W).

Figure 7.2(b) shows a rearranged sequencing of the same activities with the objective of minimising iteration, in line with most published work in this area. Minimising iteration is achieved when it is not necessary to repeat any activity upstream due to availability of data downstream. This is tantamount to saying that all data dependencies should lie below the leading diagonal.

	A	B	C	D	E	F	G	H	I	J	K	L
A	10								S			
B	S	15				W	M	S	S	M		S
C			20			S			S			
D	S			30	S			M	S			S
E				M	10			S	S			W
F						12			S	S		
G	S					S	15	W	S	M		S
H		W		S	M			10				
I									20			
J						S				10		
K						S					5	
L	M				S				S	M	M	20

Figure 7.2a *Input Matrix*

	I	A	F	J	K	E	L	D	H	G	B	C
I	20											
A	S	10										
F	S		12	S								
J			S	10								
K			S		5							
E	S					10	W	M	S			
L	S	M		M	M	S	20					
D	S	S				S	S	30	M			
H						M		S	10		W	
G	S	S	S	M			S		W	15		
B	S	S	W	M			S		S	M	15	
C	S		S									20

Figure 7.2b *The Input Matrix after re-sequencing*

As can be seen in Figure 7.2(b) most of the data dependencies are below the leading diagonal but some iteration remains, as indicated by the box enclosing activities E, L, D, H, G and B.

Although most published work stress the over-riding importance of iteration reduction, it is obvious from Figure 7.2(b) that the large number of data dependencies immediately below the leading diagonal does not provide much opportunity for concurrent pursuance of many activities, even if resources were available to do this.

A Multiple Criteria Schedule

Figure 7.3(a) shows a simple sequence of four activities, optimised on the basis of minimum iteration as there are no dependencies above the leading diagonal. The numbers on the diagonal represent activity durations in appropriate time units. The Gantt chart corresponding to this sequence could be represented as shown in Figure 7.3(b). Due to the weak data dependency between activities B and C it has been assumed that C can begin before B is complete. The degree of overlap obviously depends on domain specific characteristics.

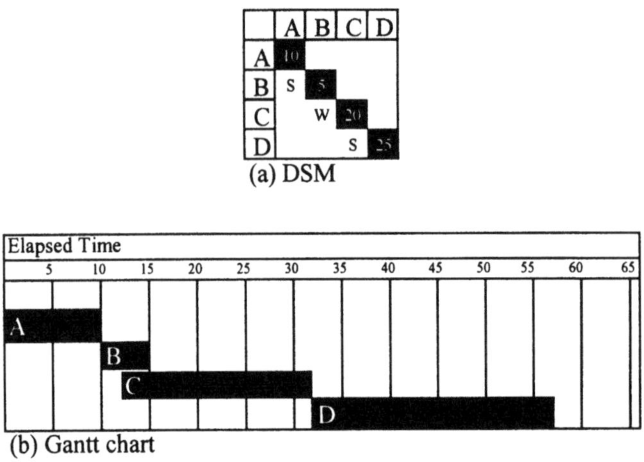

Figure 7.3 *Sequence optimised for minimum iteration*

Figure 7.4(a) illustrates the solution obtained on the basis of the twin objectives of minimum iteration and maximum concurrency. It is clear how the use of the combined objectives leads to a different solution, even for this very simple example, where a small amount of iteration may be required. This is because activity C is executed with an estimate of the data that is subsequently obtained from activity B. After completing B it may be necessary to repeat activity C using the refined data from B as the estimated data may not have been sufficiently accurate. If a full iteration is necessary the Gantt chart of Figure 7.4(c) applies. If the estimated and actual values of data obtained from B are close enough, activity C may not need to be re-executed and the result would be as shown in Figure 7.4(b).

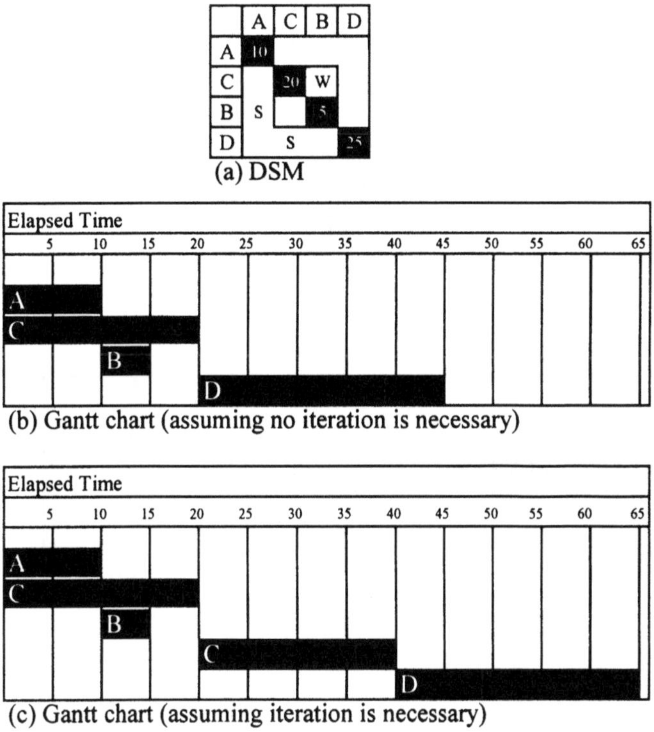

Figure 7.4 *Sequence optimised for minimum iteration and maximum concurrency*

The values of elapsed time demonstrate what this form of scheduling is all about. If an accurate estimate can be made of the data from activity B for activity C to use then the lead time for this product development process may reduce from 57 time units as in Figure 7.3(b) to 45 time units as in Figure 7.4(b). This reduction in elapsed time is at the expense of an increased bunching of resource usage as the Gantt chart shows. But this is only to be expected and is a natural corollary of concurrency.

The increased reliability of the data available may be due to information re-use from a related past project, for example. If, however, this accurate estimate cannot be made then iteration leads to considerable penalties as Figure 7.4(c) shows.

As a limiting study it is clearly instructive to examine the influence of using the single-objective of maximising concurrency.

It is clear from Figure 7.5(a) that a reversal of the activity sequence of Figure 7.3(a) is obtained. This allows concurrent execution of all activities but as there are dependencies above the diagonal, iterations would be necessary. Figure 7.5(b) and (c) show the elapsed times depending on the amount of iteration necessary. As common sense would suggest the elapsed times depends on the cycles of iterations performed. It would be observed that the total accumulated time for all activities would be quite large as well.

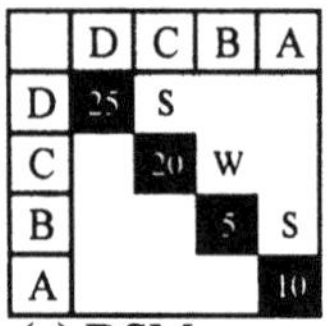

(a) DSM

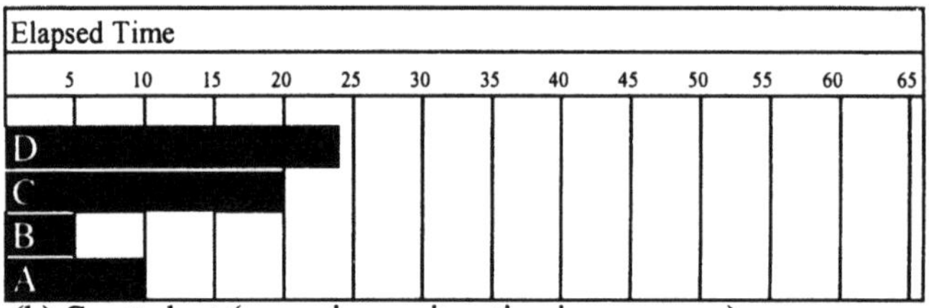

(b) Gantt chart (assuming no iteration is necessary)

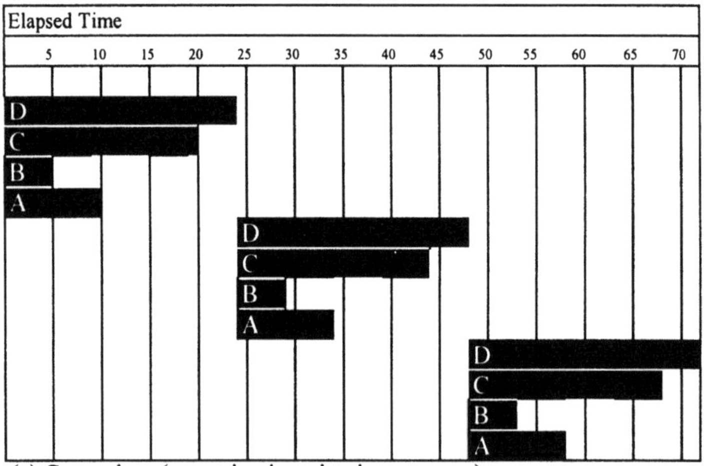

(c) Gantt chart (assuming iteration is necessary)

Figure 7.5 *Sequence optimised for maximum concurrency*

7.2.2 Multicriteria robust optimisation under uncertainty of catamarans from a seakeeping point of view

Multiple criteria decision problems can arise because of several reasons. The first study above indicates one common source; namely the presence of clearly discernible multiple goals of the design process. An alternative source of such problems is the understandable need for designs to operate over a range of different conditions with variable likelihoods associated with such conditions. This is the domain of robust optimisation, where the aim of the designer is to design artefacts that are robust with respect to the uncertain operating conditions.

Traditionally designers have attempted to achieve robust design by increasing margins but this has been of limited success. Other standard techniques used include sensitivity analysis and parametric programming, but these only provide indications of risk of sub-optimality or feasibility without providing any means of controlling them, except through a trial and error process that ultimately undermines the value of systematic sensitivity studies in the first place. The need to actively consider stochastic effects in engineering design has led progressively to the use of a large number of approaches. These have included stochastic optimisation methods like chance constrained techniques and fuzzy optimisation allied with the more traditional approaches of minimum regret. An insight that emerged from all considerations of "off-design" performance is the intimate relation between optimality and robustness, in that highly tuned optimal designs are only optimal for a narrow set of conditions and run the risk of being severely sub-optimal when those operating conditions change. It is clear that an optimisation approach that takes explicit account of the range of scenarios over which a design has to operate would be helpful in this area of work.

The aim of this example case study is to outline such an optimisation methodology for engineering design under uncertainty based on scenario decomposition and Taguchi's robust design approach. The aim of robust design is to minimise the influence of uncertain or uncontrollable parameters by minimising the variability in the response or performance of a design while keeping the mean response at some distinct target. This can be done by maximising the signal-to-noise ratio as follows:

$$\underset{x}{Max}(S/N(x) = \underset{x}{Max}\left\{-\log_{10}\left[\sum_{j=1}^{M}(f(x|z_j)-T\right]^2 \text{x}\frac{1}{M}\right\}$$

with respect to design variable x where $f(x/z_j)$ is the performance criteria or objective, M is the number of noise or uncertain factors and T is the target value for the performance criteria.

In Taguchi's method the effect of the uncontrollable or uncertain factors (noise factors) is modelled using orthogonal arrays. Taguchi's methods have been used widely to reduce the number of functional evaluations in the above process, but have not been quite so widely used in design situations where multiple measures of performance have to be considered simultaneously.

Modelling uncertainty via scenario analysis

In seakeeping analysis for advanced marine vehicles like catamarans where historical data, particular for a new design, is not available, it is necessary to employ scenario based analysis to generate the data on the basis of which robust optimisation can be performed.

The signal-to-noise based formulation of Taguchi outlined above can be generalised simply by considering a range of scenarios with associated likelihoods to represent the inherent uncertainty in operating conditions. In the case in question this uncertainty has been taken to be the directionality of the wave heading. If a collection of S scenarios are used to represent all possible headings of interest, then Taguchi's formulation can be rewritten as

$$S/N(x) = -\log_{10}\left[\sum_{s=1}^{S}(f(x/z^{s})-T)^{2} \times p_{s}\right]$$

where p_s = probability associated with scenarios s

z^s = values of uncertain parameters in scenario s.

Multiple criteria robust design formulation

In the light of comments above the signal-to-noise ratios for a single performance criterion can be extended to include multiple performance criteria by defining a ratio for each criterion. Thus, for K criteria the optimisation problem becomes the following deterministic equivalent

$$\underset{m}{Max}\left\{ S/N_{k}(x) = -\log_{10}\left[\sum_{s=1}^{S}(f_{k}(x/z^{s})-T_{k})^{2} \times p_{s}\right], k=1,2,..K \right\}$$

In contrast to single criterion optimisation, the multiple criteria formulation above can be used to derive a Pareto optimal set of non-dominated solutions.

Multiple criteria robust design of catamarans for seakeeping

The methodology outlined above is demonstrated by application to the problem of design for seakeeping of catamarans. The principal aim here is to minimise certain motion characteristics of the vessel for different wave headings, ranging between 90° (beam seas) and 180° (head seas), and to do this by choosing appropriate dimensions and hull form. This is done by modifying a parent hull form. The parent demihull has the dimensions shown in Table 7.1

Table 7.1 *The parent demihull characteristics*

Parameter	Parent Value	Parameter	Parent Value
L_p	104.00 (m)	LCB_p	45.408 (m)
B_p	9.00 (m)	LCF_p	43.306 (m)
T_p	4.50 (m)	Cwp_p	0.758
Hs_p	31.00 (m)	CB	0.397
∇	1672.55 (m^3)	Speed	40.0 Knots

where L_p, B_p, T_p, H_{sp} represent the length, breadth, draught and separation distance of demi-hulls respectively for the parent form. The parameters LCB $_p$ and LCF $_p$ represent the longitudinal position of the centre of buoyancy and centre of floatation (centroid of waterplane) with respect to the after end of the craft, and CW $_{p_p}$ is a measure of the area of the waterplane. The variables ∇ and C_b represent the underwater volume of the demi-hull and its block coefficient (a measure of fullness).

The seakeeping analysis uses linear strip theory and is fully described in Hearn et al[Hearn et al 1994]. The design development scheme consists of generation of variant demi-hulls through a systematic variation of some primary and secondary design variables. In this example application a set of variant designs is produced by small changes in the parent form. Four performance criteria are considered. These are the basic responses of heave amplitude, roll amplitude, pitch amplitude and the compound response relative bow motion. The last criterion is a measure combining the three basic responses, and allows the compound motions of the catamaran to be assessed, especially in oblique wave headings.

The six design variables $x = (x_1, x_2, \ldots x_6) = (L, B/T, H_s, C_{WP}, LCB, LCF)$ as defined in Table 7.1 are considered in two groups. The primary design variables are length (L), breadth to draught ratio (B/T) and the separation between the demihulls of the catamaran(H_s). The remaining three variables are referred to as secondary variables. The parameter z is evaluated over nine wave headings between 180° (head seas) and 90° (beam seas). In this example all headings are considered equally likely, so $p_s = 1/9$. The robust optimisation strategy thus becomes

$$\underset{x}{Max}\{\ S/N_k(x) = -\log_{10}\left[\sum_{s=1}^{S}(f_k(x/z^s) - T_k)^2 \times \frac{1}{9}\right], k = 1,2,3,4\ \}$$

Results and discussion

The non-dominated or Pareto optimal solutions are obtained by creating variant designs through a systematic variation of the six design variables and using dominance to eliminate all but the non-dominated solutions. The design variations for the six variables are as follows:

L	: 1 ± 0.1 in steps of 0.05 (i.e. ± 10% variation in steps of ± 5%)
B/T	: 1 ± 0.1 in steps of 0.05
H_S	: 1 ± 0.1 in steps of 0.05
C_{wp}	: 1 ± 0.005 in steps of 0.005
LCB	: 1 ± 0.01 in steps of 0.005
LCF	: 1 ± 0.01 in steps of 0.005

The non-dominated solutions are shown in Tables 7.2 and 7.3 for primary and secondary design variables respectively.

Table 7.2 *The non-dominated solution set for primary parameters*

	Parameters			Criteria			
No	ΔL%	ΔB/T%	ΔHs%	Heave(dB)	Pitch(dB)	Roll(dB)	RBM(dB)
1	-5.00	-10.00	-10.00	6.9372	15.3381	2.4447	6.3923
2	10.00	10.00	10.00	6.8438	19.1872	2.5865	6.8327
3	0.00	10.00	10.00	6.1819	7.9100	-0.3776	8.5016
4	5.00	5.00	10.00	6.9329	16.7476	7.4053	6.4656
5	0.00	5.00	10.00	6.1645	7.8047	5.5222	10.0961
6	10.00	0.00	10.00	6.8808	13.0745	10.8571	5.2839
7	5.00	0.00	10.00	6.9001	11.8422	8.1982	5.0751
8	-5.00	-5.00	0.00	6.9762	13.4082	5.4809	4.9013
9	-10.00	-5.00	0.00	7.2764	11.1131	4.7353	4.3594
10	0.00	-5.00	-10.00	6.1508	7.6588	8.3186	9.8139
11	-10.00	-5.00	-10.00	7.2764	11.1131	3.8911	4.9367
12	-5.00	-5.00	-10.00	6.9762	13.4082	4.7291	5.6957
13	-10.00	-10.00	-10.00	7.2411	12.9420	3.6884	5.7528
14	10.00	10.00	10.00	6.8514	17.9598	6.9071	6.1825
15	10.00	5.00	10.00	6.8908	15.8289	9.1926	5.9075
16	-10.00	-5.00	10.00	7.2764	11.1131	5.0582	3.7533
17	10.00	-5.00	10.00	6.6281	9.9164	11.4248	3.8082
18	-10.00	-10.00	0.00	7.2411	12.9420	3.9183	5.2209

Table 7.3 *The non-dominated solution set for secondary parameters*

No	Parameters			Criteria			
	ΔLCF%	ΔLCB%	ΔC_{WP}%	Heave(dB)	Pitch(dB)	Roll(dB)	RBM(dB)
1	-0.50	-0.50	0.50	5.6491	8.3388	12.4637	6.2067
2	1.00	1.00	0.50	6.4032	13.7203	-0.9263	6.8465
3	1.00	0.50	0.00	6.8449	18.5921	-5.6163	6.4729
4	-0.50	0.00	0.00	6.8450	1.6175	-4.2446	4.6551
5	-1.00	-0.50	0.00	4.7441	17.2295	-0.0829	5.7596
6	-0.50	0.00	0.50	7.0025	15.4657	-4.7494	4.0799
7	-1.00	-0.50	0.50	5.7277	15.1767	-1.8245	6.0915
8	-1.00	-1.00	0.50	5.5608	8.3986	11.8910	6.1210
9	-1.00	-0.50	-0.50	7.1489	0.2656	3.0745	2.1347
10	0.00	0.00	-0.50	6.5893	1.0935	-13.7769	6.5384
11	0.50	1.00	-0.50	6.3447	8.7142	-10.0427	7.0012
12	0.00	1.00	-0.50	6.4304	2.4858	-15.3227	7.1468
13	-0.50	1.00	-0.50	6.5185	-13.1415	-11.4519	7.1560
14	-1.00	1.00	-0.50	6.6402	-2.1119	-10.8835	6.9273
15	0.00	0.50	-0.50	6.2832	5.9648	-12.4176	7.0203
16	-0.50	0.50	-0.50	6.3349	-3.6374	-8.7489	7.1767
17	-1.00	0.50	-0.50	6.4159	-6.3352	-8.5134	7.0563
18	0.50	0.00	-0.50	2.6253	18.8067	-8.1013	5.8202
19	-0.50	0.00	-0.50	6.2545	0.1107	-3.3453	6.8935
20	-1.00	0.00	-0.50	6.2213	-9.6398	-5.4956	7.0530
21	0.50	1.00	0.00	6.3434	8.2204	-24.3333	7.0613
22	0.00	1.00	0.00	6.4110	0.3588	-12.6025	7.2035
23	-0.50	1.00	0.00	6.4983	-14.2487	-10.7109	7.1078
24	-1.00	1.00	0.00	6.5877	0.0188	-10.1876	6.8749
25	0.50	0.50	0.00	6.4365	12.8402	-4.2778	6.8298
26	0.00	0.50	0.00	6.3132	1.6242	-9.5187	7.1322
27	-0.50	0.50	0.00	6.3301	-11.2098	-8.2591	7.1647
28	0.50	1.00	0.50	6.3551	7.0521	-23.2684	7.1193
29	0.00	1.00	0.50	6.3987	-4.7251	-11.2389	7.2327
30	-0.50	1.00	0.50	6.4741	-4.4978	-10.0175	7.0600
31	-1.00	1.00	0.50	6.5648	2.8568	-9.7968	6.6192
32	0.50	0.50	0.50	6.8397	5.7261	-13.7078	7.0830
33	0.00	0.50	0.50	6.3933	-3.5069	-8.5215	7.0251

The results show the beneficial motion characteristics can be obtained for both positive and negative changes in design variables. However, some of the detailed results are worth noting. For heave response the best solutions are (9,11,16). These results indicate that a reduction in L and B/T is beneficial, irrespective of changes in H_S. Thus the value of H_S can, within reason, be chosen on the basis of other considerations. The best solution for pitch (2,14) indicate maximum increase in all three primary design variables. In roll the two best solutions (6,17) have the maximum permissible increases in L and H_S with either no or negative increase in B/T. The principal point of interest is that conflicting changes in design can produce comparable signal-to-noise ratios.

This is evident in both sets of results. For example, the three best designs (3,5,18) from the pitch point of view in Table 7.3 show this clearly. Two of the solutions, 3 and 5, indicate moving LCF aft 1% of length and LCB aft 0.5% or moving them forward the same amounts. The best solution, which is number 18, shows quite different results. This, of course, is only to be expected as good aspects of specific types of performance are traded for more average performances across the whole spectrum of measures.

To identify the overall best solution from the non-dominated sets above the reference point method [Lewandowski and Wierzbicki 1989] is used. Table 7.4 shows the best compromise results

Table 7.4 *The best compromise robust solutions*

	Parameters			Criteria			
Secondary parameters	ΔLCF %	ΔLCB %	ΔC_{WP} %	Heave (dB)	Pitch (dB)	Roll (dB)	RBM (dB)
Reference point				7.1489	18.8067	12.4637	7.1767
Best compromise solution	-1.00	-1.00	0.50	5.5608	8.3986	11.8910	6.1210
	Parameters			Criteria			
Primary parameters	ΔL %	ΔB/T %	ΔHs %	Heave (dB)	Pitch (dB)	Roll (dB)	RBM (dB)
Reference point				7.2764	19.1872	11.4248	10.0961
Best compromise solution	10.00	5.00	10.00	6.8908	15.8289	9.1926	5.9075

Interestingly, the best compromise solution agrees with earlier cited findings based on head and bow seas only. The results therefore indicate that it is possible to identify one set of primary and secondary changes to the parent hull to give benefits across all aspects of performance. This gives greater confidence in the robustness of the suggested optimal solution.

7.3 Concluding Comments

The entire domain of MCDM has moved a long way since its early beginnings and has exhibited some distinct phases. This is largely in harmony with all comparable areas of human endeavour. It is necessary first to classify problems to find common patterns. These patterns are now quite clear in MCDM and have been so for some time. The next step is the development of methods and in MCDM this phase is in full swing with methodology being created at a very rapid rate as each answer leads to even more questions. Whilst this is a natural and healthy state of affairs it is important that all methodologies be examined in the light of their underlying assumptions and their data requirements to see whether their use would be justifiable in the particular context. The third and in many respects the natural final stage of development of any methodology is its incorporation in day-to-day decision making. This is not to say that developments come to an end or the classification exercise is no longer active. The first two phases continue to be revisited and worked on, but the discipline of actual implementation place all developments in context.

The essential features of any decision-making tool are that it should be transparent and efficient in use. While these two objectives are often essentially met simultaneously in the same approach, this is unfortunately not always the case. The choice of appropriate methodology, therefore, is an important consideration. From the point of view of the engineering designer it is important to note that the final solution chosen depends on what is technically feasible and simultaneously in harmony with the designer's or DM's priority ordering of requirements. But even these two considerations are not independent as given the hierarchy of designer requirements what is technically feasible may indeed depend on the perceived priorities of the designer. Given the large choice of methodology the designer has the opportunity to decide on a range of approaches before settling on a design. If this book has contributed in clarifying that choice it would have amply served its purpose.

So much for the past and the present. But what can one look forward to in the future? It is clear that one of the most important areas of work is to examine and codify how DMs make decisions, and to bring these observations to bear on the application of MCDM methods. The smaller the gap between theory and practice the greater is the confidence attached to the worth of the solutions obtained thereby. The increased development of evolutionary computation techniques has opened up new opportunities for the intelligent exploration of multidimensional solution spaces. This allows the realistic implementation of multiple criteria techniques into more complex domains. This is another area of endeavour that is likely to see further developments, particularly in terms of parallel algorithms, as the computational burden becomes more onerous.

Complex environments also demand more subtle expressions of preference and any judgements made over such environments are also accompanied by considerations of uncertainty and risk, as the operating environment and the performance of a design in it are both likely to be variable. Formal methods of taking such uncertainty realistically into account will be increasingly required if robust designs are to be arrived at.

It may be expected that some of the more interesting developments over the next few years will be in some of the above areas of investigation. It may also be expected that some unexpected developments will also take place, thereby opening up new ways of asking questions and receiving answers. That too is a feature of all developments like those in MCDM where each generation of methodologies build on those that have gone before. Therefore, whatever these emerging advances may be, in a sense they would also have been anticipated.

References

Andrews D 1981 Creative Ship Design. *Transactions of Royal Institution of Naval Architects,* Vol.123, 447-471.

Austin S, Baldwin A, Newton A 1996 A data flow model to plan and manage the building design process. *Journal of Engineering Design,* Vol.7 No.1 3-25.

Belton V, Gal T 1981 On a short-coming of Saaty's method of analytic hierarchy, *Omega* Vol.11, No.3, 228-230.

Bouyssou D 1990. Building criteria : a prerequisite for MCDA, Bana e Costa CA (ed.) *Readings in Multiple Criteria Decision Aid,* Springer Verlag, Berlin.

Chankong V, Haimes Y Y 1991 *Multiobjective Decision Making: Theory and Methodology,* North-Holland, New York.

Charnes A, Cooper W W 1977 Goal programming and multiple objective optimisation – Part 1, *European Journal of Operational Research,* 1, 39-54.

Dyer J S 1972 Interactive goal programming, *Management Science,* 19/1, 62-70.

Evans G W 1984 An overview of techniques for solving multiobjective mathematical programme. *Management Science,* Vol.30, No.11.

Fonseca C M, Fleming P J, 1993 Genetic algorithms for multiobjective optimisation: formulation, discussion and generalisation, *Fifth International Conference on Genetic Algorithms,* Morgan Kaufman, 416-423.

FRONTIER 1995 Open system for collaborative design optimisation using Pareto frontiers, ESPRIT Project No.20082.

Gabbert P, Brown D E 1989 Knowledge-based computer-aided design of material handling systems, *IEEE Transactions on Systems, Man & Cybernetics,* Vol.19, No.2, 188-196.

Gebala D A, Eppinger S D 1991 Methods for analysing design procedures. *Proceedings of the ASME Design Theory & Methodology (DTM'91)* Conference, 227-233.

Geoffrion A M, Dyer J S, Feinberg A 1972. An interactive approach for multicriteria optimisation, with an application to the operation of an academic department, *Management Science,* 19/4, 357-368.

Gero J, Louis S J 1995 Improving Pareto optimal designs using genetic algorithms, *Microcomputers in Civil Engineering*, Vol.10, 239-247.

Goldberg D E 1989 *Genetic Algorithms in Search, Optimization and Machine Learning*, Addison Wesley.

Hajela P, Lin C Y 1992 Genetic search strategies in multicriterion optimal design, *Structural Optimization*, 99-107.

Horn J, Nafpliotis 1993 Multiobjective optimisation using the niched Pareto genetic algorithm, IlliGAL Report No.93005, Illinois Genetic Algorithms Laboratory, University of Illinois at Urbana Champaign, Il, USA.

Hwang C L, Masud A S 1979 *Multiple Objective Decision Making Methods and Applications, A State-of-the-Art Survey*, Springer-Verlag, Berlin.

Hwang C L, Yoon K 1981 *Multiple Attribute Decision Making Methods and Applications, A State-of-the-Art Survey*, Springer-Verlag, Berlin.

Islei G, Lockett A G 1988 Judgemental modelling based on geometric least squares. *European Journal of Operations Research*, Vol.36, 27-35.

Jacquet-Lagreze E, Siskos J 1982 Assessing a set of additive utility functions for multicriteria decision making, the UTA method, *European Journal of Operational Research*, 10, 151-164.

Johnson C R, Beine W B, Wang T J 1979 Right-left asymmetry in an eigenvector ranking procedure, *Journal of Mathematical Psychology*, No.1 61-64.

Keeney R L, Raiffa H 1976 *Decisions with Multiple Objectives: Preferences and Value Trade-Offs*, Wiley, New York.

Kursawe F 1991 A variant of evolution strategies for vector optimisation, *Parallel Problem Solving for Nature*, Vol.496 of Lecture Notes in Computer Science, 193-197.

Kusiak A, Park K 1990 Concurrent engineering: decomposition & scheduling of design activities, *International Journal of Production Research*, Vol.28, No.10, 1883-1900.

Lewandowski A, Wierzbicki A P 1989 *Aspiration Based Decision Support Systems*, Springer-Verlag.

Lightner M R, Director S W 1981 Multiple criterion optimisation for the design of electronic circuits. *IEEE Transactions on Circuits and Systems*, 28/3 169-179.

Luenberger D G 1973 *Introduction to Linear and Nonlinear Programming*, Addison-Wesley, Reading, Massachusetts, USA.

Meldrum P F, Sen P, Yang J B 1990 The analytic hierarchy process: the problem of right-left eigenvector asymmetry, Research Report, Engineering Design Centre, University of Newcastle.

Minoux M 1986 *Mathematical Programming – Theory and Algorithms*, John Wiley & Sons, Chichester.

Murata T, Ishibuchi H 1994 Performance of genetic algorithms for flowshop scheduling problems, *Proceedings of IEEE Conference on Evolutionary Computation,* 289-294.

Nijkamp P, van Delft A 1977 *Multi-criteria Analysis and Regional Decision-making.* Martinus Nijhof Social Science Division, Leiden, The Netherlands.

Ozernoy V M 1987 A framework for choosing the most appropriate discrete alternative multiple decision making method in decision support systems and expert systems. *Toward Interactive & Intelligent Decision Support systems* (LNEMS 286), Springer-Verlag, Vol.2, 56.

Pahl G, Beitz W 1977 *Engineering Design,* a *Systematic Approach,* K.Wallace (ed.). The Design Council, London and Springer Verlag, Berlin Heidelberg.

Press W H, Teukolsky S A, Vetterling W T, Flannery B 1992 *Numerical Recipes in C The Art of Scientific Computing,* Cambridge University Press.

Richardson J T, Palmer M R, Leipins G, Hilliard M 1989 Some guidelines for genetic algorithms with penalty functions, *Proceedings of the Third International Conference on Genetic Algorithms,* Morgan Kaufman, 191-197.

Ritzel B J, Eheart J W, Ranjithan 1994 Using genetic algorithms to solve a multiple objective groundwater pollution containment problem, *Water Resources Research,* Vol.30, No.5, 1589-1603.

Rogers J 1997 Reducing design cycle time and cost through process re-sequencing, *Proceedings of the International Conference on Engineering Design* (ICED'97), Vol.1, 193-198.

Saaty T L 1988 *The Analytic Hierarchy Process.* University of Pittsburg, USA.

Schaffer J D 1985 Multiple objective optimisation with vector evaluated genetic algorithms, *Proceedings of the First International Conference on Genetic Algorithms and their Applications,* Lawrence Erlbaum Associates, Hillsdale NJ, 93-100.

Sen P, Fulford K, Buxton I L et al 1988 Efficient Ship Programme Main Report, Integrated Design Project, DTI Contract No.RD/2101/9/037.

Sen P, Yang J B 1993a A multiple criteria decision support environment for engineering design. Proceedings of 9th *International Conference on Engineering Design,* The Hague, The Netherlands, 465-472.

Sen P, Yang, J B, Meldrum P F 1993b Interactive trade-off analysis in multiple criteria preliminary design of a semi-submersible. *Proceedings of ICMES'93 Marine System Design and Operation,* Technical Univ. of Hamburg-Harburg, Germany, 10.9-10.19.

Sen P, Yang J B 1994a Design decision making based upon multiple attribute evaluation and minimal preference information. *Mathematical and Computer Modelling,* Vol.20, No.3, 107-124.

Sen P, Yang J B 1994b Combining objective and subjective factors in multiple criteria *marine design. Proceedings of 5th International Marine Design Conference & Summer Meeting of the German Society of Naval Architects*, Delft, The Netherlands, 505-519.

Sen P, Yang J B 1994c Preference modelling in multiple criteria design decisions, *Proceedings of the Conference on Informing Technologies to Support Engineering Decision Making*, London UK, 133-142.

Sen P, Yang J B 1995a Multiple criteria decision making in design selection and synthesis, *Journal of Engineering Design*, Vol.6, No.3, 207-230.

Sen P, Yang J B 1995b An investigation into the influence of preference modelling in ship design with multiple objectives, *Proceedings of the Sixth International Symposium on Practical Design of Ships and Mobile Units*, Seoul, Korea, September, 17-22.

Simon H 1981 The Sciences of the Artificial, The M.I.T.Press, Cambridge, Massachusetts.

South M C, Wetherill G B, Tham M T 1993 Hitch-Hiker's guide to genetic algorithms, *Journal of Applied Statistics*, Vol.50, No.1 153-174.

Srinivas N, Deb K 1995 Multiobjective optimization using non dominated sorting in genetic algorithms, *Evolutionary Computation*, Vol.2, No.3, 221-248.

Steuer R E, Choo E U 1983 An interactive weighted Tchebycheff procedure for multiple objective programming. *Mathematical Programming*, 26, 326-344.

Steward D V 1981 *Systems Analysis & Management*, Petrocelli Books.

Stewart T J 1987 An interactive multiple objective linear programming method based on piecewise linear additive value functions. *IEE Transactions on Systems, Man &* Cybernetics, 17, 799-805.

Stewart T J 1992 A critical survey on the status of multiple criteria decision making theory and practice. *Omega* 20/5-6, 569-586.

Suh N P 1990 The Principles of Design, Oxford University Press, New York.

Teghem J, Delhaye C, Kunsch P L 1989 An interactive decision support system (IDSS) for multicriteria decision aid. *Mathematical and Computer Modelling*, Vol.12, No.10/11, 1311-1320.

Vanderpooten D, Vincke P 1989 Description and analysis of some representative interactive multicriteria procedures. *Mathematical and Computer Modelling*, 12, 1221-1238.

White III C C, Sage A P, Dozono S 1984 A model of mutliattribute decision making and trade-off weight determination under uncertainty. *IEEE Transactions on Systems, Man & Cybernetics*, SMC-14, No.2, 223-229.

Yang J B, Chen C, Zhang Z J 1988 The interactive decomposition method for multiobjective linear programming and its applications. *Information and Decision Technologies*, Vol.14, No.3, 275-288.

Yang J B, Chen C, Zhang Z J 1990 The interactive step trade-off method (ISTM) for multiobjective optimisation. *IEEE Transactions on Systems, Man & Cybernetics*, Vol.20, No.3, 688-695.

Yang J B 1992a A decision support software sub-system for multiple attribute decision making: C source code and applications, Research Report, Engineering Design Centre, University of Newcastle.

Yang J B. 1992b A decision support software sub-system for multiple objective decision making: C and FORTRAN source code and applications. Research Report, Engineering Design Centre, University of Newcastle.

Yang J B 1992c An integrated multiple criteria decision support system for engineering design. Research Report, Engineering Design Centre, University of Newcastle. (60 pages).

Yang J B, Sen 1993a P A hierarchical evaluation process for multiple attribute design selection with uncertainty. *Industrial and Engineering Applications of Artificial Intelligence and Expert Systems* (IEA/AIE-93), Chung, PWH, Lovegrove, G.and Ali, M. (eds.) Gordon and Breach Science Publishers, 484-493.

Yang J B, Sen P 1993b An interactive MODM method for design synthesis with assessment and optimisation of local utility functions. *PEDC Seminar on Adaptive Search and Optimisation in Engineering Design*, Plymouth University, UK.

Yang J B, Singh M G 1994a An evidential reasoning approach for multiple attribute decision making with uncertainty. *IEEE Transactions on Systems, Man & Cybernetics*, Vol.24, No.1, 1-18.

Yang J B, Sen P 1994b A general multi-level evaluation process for hybrid MADM with uncertainty. *IEEE Transactions on Systems, Man & Cybernetics*, Vol.24, No.10, 1458-1473.

Yang J B, Sen P 1994e An artificial neural network approach for nonlinear optimisation with discrete design variables. Lecture Notes in Control and Information Sciences, Henry J. and Yvon J-P.(eds.), Springer-Verlag, London, 671-770.

Yang J B, Sen P 1994d Evidential reasoning based hierarchical analysis for design selection of ship retro-fit options. *Artificial Intelligence in Design '94*, Gero, J S. and Sudweeks F. (eds.). Kluwer Academic Publishers,The Netherlands, 327-344.

Yang J B, Sen P 1994e. Multiple objective design optimisation by estimating local utility functions. *Proceedings of the 20th ASME Design Automation Conference*, Minneapolis, Minnesota, USA.

Yang J B, Sen P 1996a Preference modelling by estimating local utility functions for multiobjective optimisation. *European Journal of Operational Research*, Vol.95, 115-138.

Yang J B, Sen P 1996b Interactive trade-off analysis and preference modelling for preliminary multiobjective ship design. *Systems Analysis, Modelling and Simulation*, Vol.26, 25-55.

Yang J B, Sen P, Meldrum P 1997 Multiple attribute evaluation in engineering decision support using limited compensation and reference designs. *Information and Systems Engineering,* Vol.2, Nos.3&4, 159-181.

Zhang Z J, Yang J B, Xu DdL 1990 A hierarchical analysis model for multiobjective decision making. Analysis, Design and Evaluation of Man-Machine System 1989. *Selected Papers from 4th IFAC/IFIP/IFORS/IEA Conference,* Xian, P.R.China, Oxford UK, Pergamon, 13-18.

TOPIC INDEX

MIX
Papier aus verantwortungsvollen Quellen
Paper from responsible sources
FSC® C105338

If you have any concerns about our products,
you can contact us on
ProductSafety@springernature.com

In case Publisher is established outside the EU,
the EU authorized representative is:
Springer Nature Customer Service Center GmbH
Europaplatz 3, 69115 Heidelberg, Germany

Printed by Libri Plureos GmbH
in Hamburg, Germany